Teaching and Learning Mathematics in the 1990s

1990 Yearbook

Thomas J. Cooney
1990 Yearbook Editor
University of Georgia

Christian R. Hirsch
General Yearbook Editor
Western Michigan University

National Council of Teachers of Mathematics

ISSN: 0077-4103
ISBN: 0-87353-285-6

Teaching and learning mathematics in the 1990s / Thomas J. Cooney, 1990 yearbook editor, Christian R. Hirsch, general yearbook editor.
p. cm. — (Yearbook, ISSN 0077-4103 ; 1990)
Includes bibliographical references.
ISBN 0-87353-285-6
1. Mathematics—Study and teaching. I. Cooney, Thomas J. II. Hirsch, Christian R. III. National Council of Teachers of Mathematics. IV. Series: Yearbook (National Council of Teacher of Mathematics) ; 1990.
QA1.N3 1990
[QA12]
510 s—dc510/ .71 —dc20
89-14410
CIP

The publications of the National Council of Teachers of Mathematics present a variety of viewpoints. The views expressed or implied in this publication, unless otherwise noted, should not be interpreted as official positions of the Council.

Printed in the United States of America

Contents

Part 3: The Role of Assessment in Teaching and Learning

Part 4: Cultural Factors in Teaching and Learning

Part 5: Contextual Factors in Teaching and Learning

Part 6: Implications of Technology for Teaching and Learning

Preface

During the 1980s this series of NCTM yearbooks focused on curricular issues and on the teaching of specific content. As we enter a new decade, the Educational Materials Committee has wisely chosen to devote the 1990 Yearbook to the people of mathematics education—teachers and students—and their changing roles in the face of the calls for reform in the 1990s. Perhaps never before have so many mathematics educators been faced with such diverse issues.

This yearbook is organized into seven sections, as indicated in the Table of Contents, to address many of these issues. Articles 1–5, which constitute the first section, focus on the relationship between research and practice and suggest a perspective based on the belief that mathematical learning consists of students constructing mathematical concepts and procedures. This perspective is in sharp contrast to the notion that mathematics is a "received" body of knowledge in which the teacher's role is to "transmit" and the students' role is to "receive." The "constructivist" perspective of this section is shown in practical settings and offers suggestions for teaching from such a perspective.

Articles 6–12 form the second section of the yearbook and offer suggestions for effective methods of teaching mathematics. The orientation of these articles and their suggestions for teaching are consistent with NCTM's *Curriculum and Evaluation Standards for School Mathematics*. The use of small-group instruction, suggestions for effective use of class time, an emphasis on communication and writing as vehicles for instruction, the importance of problem posing, and the legendary problem of motivating students are addressed in these articles.

But reform in the teaching of mathematics is not likely to occur without an accompanying reform in the ways we assess students' understanding of mathematics. How we can broaden our horizons in assessing students' mathematical understanding is the focus of articles 13 and 14, which constitute the third section of the yearbook.

Forces for reform have been fueled not only by a movement that suggests that a paradigm shift is needed on how mathematics should be developed and learning assessed but also by the issue of how to promote numeracy for all our citizens. The fourth section, articles 15–20, addresses this issue in the face of changing demographics in the United States. Of particular importance is the question of how we can increase the participation of minority groups and women in the pursuit of mathematics and the concomitant issue of how cultural diversity influences what mathematics is taught and how it is learned.

Any attempt to address reform must necessarily consider the role students play in shaping classroom events and the importance of considering the contexts of mathematical applications. These important, but often neglected, considerations are addressed in the fifth section, articles 21 and 22.

Another force for reform that is addressed in the sixth section, articles 23–26, is that of technology, its increased role in our society, and the various types of technology readily available for classroom use. Clearly, changes in what and how mathematics should be taught will be the basis for extensive thought and analysis in the coming decade in light of technological advances. Perhaps never before has mathematics education entered such unchartered waters with respect to the teaching and learning of mathematics as a consequence of this advancement.

The last section, articles 27 and 28, takes the point of view that if reform is to be realized, then we must empower teachers as decision makers who have the responsibility and the latitude to shape classroom events in accordance with the best of what is known about the teaching and learning of mathematics. It follows that teachers' roles must be reconceptualized as they are given increased professional responsibility.

Finally, I would like to thank a number of people for their contributions to this yearbook. Thanks are extended to the authors for their willingness to share their insights with others. For their boundless energy and dedication in helping to shape this yearbook, I express thanks and gratitude to the members of the 1990 Yearbook Editorial Panel:

Edward J. Davis	University of Georgia
Donald J. Dessart	University of Tennessee
Christian R. Hirsch	Western Michigan University
David R. Johnson	Nicolet High School, Milwaukee (Glendale)
Miriam A. Leiva	University of North Carolina at Charlotte

I would particularly like to thank the general editor, Chris Hirsch, for continually cudgeling his brain for ways to improve the yearbook. His support and extensive input are greatly appreciated. And certainly thanks are extended to the NCTM staff for their constant vigilance in making this yearbook the best possible one. We all hope that readers will find it thought-provoking and useful as they grapple with the many issues facing us in the coming decade.

THOMAS J. COONEY
1990 Yearbook Editor

Part 1

1

Contributions of Research to Practice: Applying Findings, Methods, and Perspectives

Edward A. Silver

THIS article is about the relationship between educational practice and educational research on the teaching and learning of school mathematics. The general relationship between research and practice in education has been discussed extensively over at least the past fifty years, and recent debates attest to the fact that this relationship remains a topic of importance and interest in both communities.

Educational research suffers from a fairly widespread belief that some day a research study will provide the ultimate answer to each of the most pressing questions of educational practice. The belief that we can identify educationally important targets, conduct research studies aimed at those targets, and supply unequivocal answers to our questions or cures for our educational problems may be related to our experience with our research fields, such as medicine. Perhaps we view educational research metaphorically as a search for a magic cure. This belief leads to unrealistic expectations. Not only are important questions with educational significance often too complex to be settled in this way, but the world of educational realities is also often too uncontrollable. Implicit in this article is the view that we need to alter our adherence to this belief and the search for a magic cure. A better metaphor for thinking about the influence of educational research might be "osmosis"—the general permeation of the field of educational practice by ideas and constructs from the field of educational research, and vice versa. In this view, ideas and perspectives emanating from research and from practice are viewed as merging to help form the zeitgeist in the mathematics education community at any given time.

The preparation of this article was supported, in part, by the National Science Foundation through Grant No. MDR-8850580. The opinions expressed are those of the author and not necessarily those of the Foundation.

This article is based on the premise that several different aspects of research on the teaching and learning of mathematics have the potential for making substantial contributions to educational practice. The most obvious, but not the only, aspect of research that has potential applicability to practice is research *findings*. The utility of the specific findings of a single research study is probably somewhat overestimated by educational practitioners, at least in part because of the prevailing search for the magic cure. Nevertheless, the results of systematic programs of research can develop cumulative results that lend themselves to substantive interpretation and important implications for practice. Less obvious, but potentially quite influential and helpful, are the *methods* used in research and the *theoretical constructs* or general theoretical *perspectives* used to frame a research endeavor or to explain its findings. We shall discuss each of these in general and give specific examples drawn from recent research on mathematics teaching or learning.

USING RESEARCH FINDINGS

Significant findings from programmatic research in many areas of mathematics education have accumulated, and many topics would be suitable for discussion. For example, systematic attention has been paid to the areas of problem solving (Charles and Silver 1988; Schoenfeld 1985; Silver 1985), rational and decimal number learning (Hiebert and Behr 1988), elementary algebra (Wagner and Kieran 1989), early number learning and arithmetic problem solving (Carpenter, Moser, and Romberg 1982), and mathematics teaching (Grouws, Cooney, and Jones 1988). However, since research summaries already exist for these areas, our focus here will instead be on some of the findings from two lines of research inquiry that are somewhat less known in the mathematics education community but that appear to be applicable to instructional practice in our field.

Among the important decisions to be made in planning mathematics instruction are choices concerning the number and variety of examples to be included for the learner. The importance of practice with examples in the learning of mathematics has long been recognized, but contemporary research suggests a somewhat more refined view of the role that examples may play in the acquisition of expertise, and the research discussed here suggests that it might be wise to give some consideration to providing alternative forms of practice, such as worked-out examples.

Worked-out examples, such as those provided in many study guides for college mathematics and science courses, have the virtue that more of them can be accommodated per unit of instructional time than examples requiring the learner to generate an original solution. Some recent work on learning mathematical procedures from worked-out examples suggests that learners can profit from experience with them and sheds some light on the mechanisms that may underlie successful learning from examples. Sweller and Cooper (1985) studied the use of worked-out examples as a substitute for conventional exercises in learning to solve algebraic equations. They found that students who studied a set of correctly worked examples made fewer errors on a posttest than students who were given conventional problems in the form of exercises to solve. They suggest that students studying worked examples directly process the relationship between the given initial state and the steps required in order to achieve the goal state, thereby increasing the probability of acquiring both a knowledge of the relationships among different items of knowledge and a mastery of processes that exploit these relationships. Individual differences noted in the Sweller and Cooper study suggested that students' motivation may be an important mediating factor. If students adopt a passive role when learning from worked examples, they are less likely to achieve success.

In closely related work, Zhu and Simon (1987) compared the performance of students who learned to factor quadratic expressions through conventional instruction with that of students who learned by studying a carefully constructed set that combined worked-out examples and conventional exercises. They reported that the students who studied worked-out examples were at least as successful on all performance measures as the students who had learned by conventional methods. Moreover, evidence obtained from interviews indicated that the students who studied the worked-out examples did not simply learn rote procedures but rather learned with understanding. According to Zhu and Simon, this successful learning from examples is due, at least in part, to the fact that the students who studied the worked-out examples were actively engaged in their learning—spending their time studying the examples and examining relationships among solution steps—rather than passively listening to a teacher's explanation.

Further insight into students' learning from worked examples is provided by Chi, Bassok, Lewis, Reimann, and Glaser (1989), who analyzed the processes used by college-level science students in studying worked examples of solutions to elementary mechanics problems. Their data indicate that successful learning from examples is characterized by the generation of explanations that refine and expand the implicit actions in the example solutions and relate these actions to scientific principles. However, unsuccessful students tend to rely heavily on the examples alone to provide the learning.

Smith and Silver (1989) recently explored the effectiveness of using worked examples in a remedial setting to help students unlearn the cancellation error in algebra. Their study demonstrated that students could make effective use of worked examples to eliminate the error from their procedural repertoire.

This program of research is just emerging, but the findings of the research on learning from worked-out examples are provocative and suggest that the use of such examples could influence mathematics instruction. If applied to procedural learning, the use of worked-out examples might provide a mechanism for efficient learning of procedures, thereby freeing valuable classroom instructional time for engagement with richer problem situations and more extensive interactions among students and teachers. If applied judiciously to certain classes of mathematics problems, worked-out examples could allow learners to become acquainted with a large number of problem solutions much more rapidly than if all the solutions were generated by the learner. However, it is clear that students differ in their ability to benefit substantially from worked-out examples. In order to make effective use of worked-out examples, the instructional program might also have to give some attention to the comprehension-monitoring skills and intellectual disposition apparently needed to profit from such instruction.

Another line of research inquiry suggests the potential utility of a different kind of activity that might be included in school mathematics instruction: problems with nonspecific rather than well-specified goals. Figure 1.1 presents several examples of goal-specific and non–goal-specific mathematics problems. These problems relate closely to the current interest in the role of open-ended investigation and problem-posing activities as components of classroom instruction (NCTM 1989). Although problem posing has not been extensively studied in research on mathematical problem solving (Kilpatrick 1987), some findings from closely related research appear to be instructionally relevant.

A series of studies contrasting students' learning from goal-specific problems (e.g., fig. 1.1, Problem 1A) with that from non–goal-specific problems (e.g., fig. 1.1, Problem 1B) has been conducted by John Sweller, an Australian educational psychologist, and his colleagues (Sweller and Levine 1982; Sweller, Mawer, and Ward 1983; Owen and Sweller 1985). Many of Sweller's studies have involved mathematics problems from the domain of geometry and trigonometry. In general, the results of these studies have indicated that students are often able to learn usable knowledge and skills more effectively and efficiently through experience with non–goal-specific problems and exercises than with more traditional goal-specific versions. According to Sweller and his colleagues, when students solve the goal-specific problems, they are likely to use general strategies that are effective for solving the specific exercises or problems but that are less effective for making connections among concepts and procedures or for organizing knowledge. In contrast, the non–goal-specific problems apparently offer opportunities for students to use strategies that make important relationships more salient,

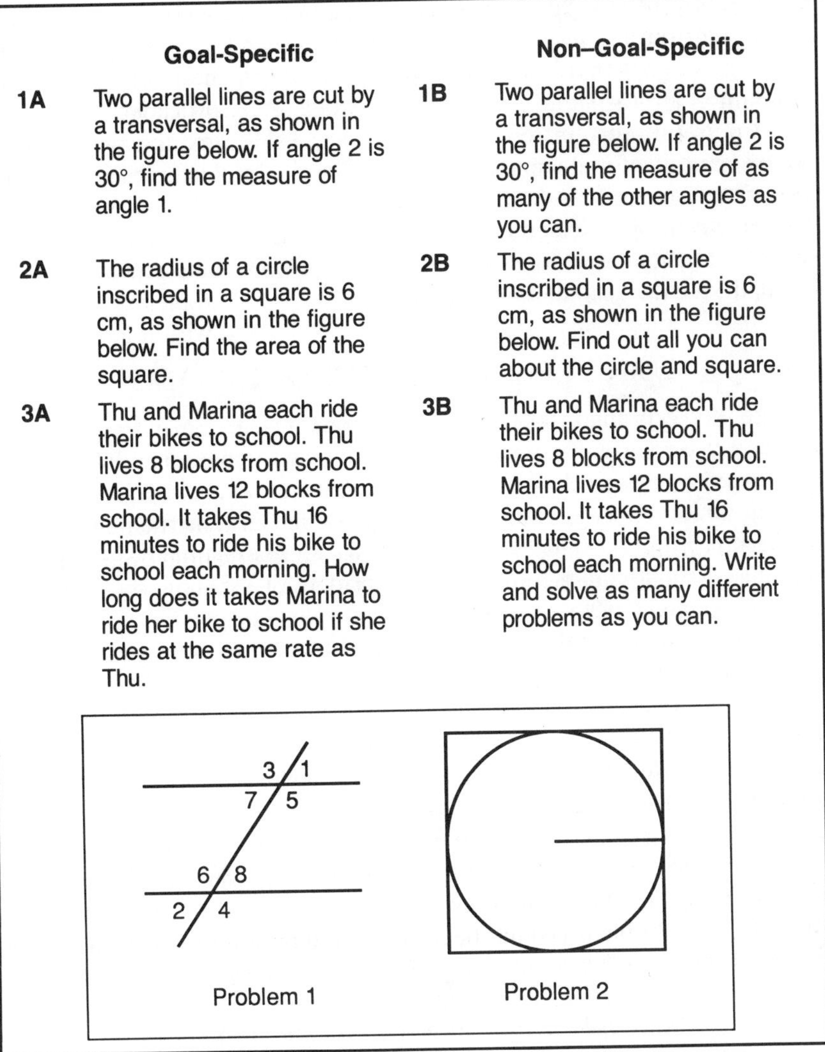

Fig. 1.1. Examples of goal-specific and non–goal-specific problems

thereby helping students to develop knowledge that is better organized and skills that are more usable.

Extrapolating from Sweller's findings, one would predict that students' long-term engagement with non–goal-specific problems and the associated problem-posing and conjecturing activities may have a dramatic, positive effect on their subsequent knowledge or problem solving. Consonant with current calls for the reform of precollege mathematics instruction (e.g., NCTM 1989), which argue that generative mathematical activities, such as problem posing and conjecturing, should be more prevalent in mathematics instruction, the findings of this program of research on non–goal-specific problems suggest that it may be wise for us to include in our instructional program at least some of these kinds of problems as occasions for students to learn useful problem-solving knowledge and skills and to engage in generative aspects of mathematical thinking, such as problem posing and conjecturing. Silver and Adams (1987) recently presented some examples of ways in which open-ended problems might be used in teaching elementary school mathematics, and Brown and Walter (1983) have provided several examples for the secondary school level.

USING RESEARCH METHODS OR TASKS

The methods and tasks used in educational research may also provide a rich source for the instructional developer or classroom teacher. Researchers often need to design clever, nonstandard tasks to assess important mathematical understandings. Therefore, the tasks and general methods used in research represent a valuable source of tasks for instructional or assessment activities for mathematics educators. Consider, for example, the non–goal-specific problems used by Sweller and his colleagues in the research discussed above. Non–goal-specific tasks like those shown in figure 1.1 are only slight modifications of those found in many textbooks, but as we have discussed, research suggests that this relatively minor modification could provide students with important opportunities for more exploration and problem solving than the typical textbook task, which usually requires the attainment of a single, well-specified goal.

Another set of tasks that might be especially useful to classroom teachers is drawn from the literature on problem similarity. A number of studies (e.g., Schoenfeld and Herrmann 1982; Silver 1979) have attempted to probe students' perceptions of similarity among mathematical problems as a clue to how students might organize problem-solving information in memory or make use of Polya's heuristic suggestion, "Think of a related problem." Many of the tasks used in these research studies—tasks in which students form groups of problems according to perceived mathematical similarity, judge the degree of similarity between pairs of problems, or choose the "most related" problems in a set—are suitable for use in instruction or assessment. Much can be learned about students' thinking through the use of these research-based tasks. Silver and Smith (1980) have described classroom uses for many of these research tasks.

In addition to specific tasks drawn from research studies, general methodological approaches might also find use in the mathematics classroom. For example, researchers have made extensive use of clinical and "think aloud" interviews in order to probe and understand the thinking of individual students. It would be a rare classroom in which a teacher would have the luxury of conducting clinical interviews with each student; yet, much might be gained from adapting this popular research approach for classroom use. Many teachers know that an occasional probing question to an individual child or group of children can be used to assess current conceptions or misunderstandings. Although the suggestion that student interviews might reap substantial dividends for mathematics teachers is not a completely new one (Weaver 1955), the extensive use of clinical interviews in recent research has taught us that much can be learned when students are invited to make their thinking and reasoning public. Moreover, it is also clear that student verbalization can not only help teachers to gain insight into the knowledge and thinking of their students but also furnish a powerful way for students to learn from each other.

USING THE THEORETICAL CONSTRUCTS AND PERSPECTIVES OF RESEARCH

Another way in which research on learning and performance can affect instructional practice is through the influence of the theoretical constructs and perspectives derived from that research. It could be argued that much of the current interest in problem solving in mathematics education is due in large part to the influence of theoretical constructs, such as heuristic processes, and theoretical perspectives, such as an orientation toward cognitive processes rather than cognitive products, that have been provided in the past few decades of research in this area. Although the influence of these theoretical constructs and perspectives on the mathematics education community is subtle and often difficult to detect, there is little doubt that such influence can be demonstrated. Another example of this subtle impact is the currently high level of interest in the role of metacognition—with respect to the importance of not only self-knowledge and personal beliefs but also self-regulation, monitoring, and evaluation—which can be traced, in large part, to the interest in metacognitive aspects of mathematical performance and problem solving in the mathematics education research community (e.g., Schoenfeld 1987; Garofalo and Lester 1985; Silver, Branca, and Adams 1980).

Another theoretical perspective that has permeated the mathematics education community is a very general form of constructivism in which it is acknowledged that students actively and personally construct their own knowledge rather than making mental copies of knowledge possessed and transmitted by teachers or textbooks. Although many would agree that a deep appreciation of the constructivist perspective has yet to take hold in the larger mathematics education community, a very general version of constructivism has certainly taken root, and its implantation can be traced directly to the influence of research—from the writings of Piaget to the work of many prominent contemporary researchers in psychology and mathematics education.

In recent years another research perspective has emerged that may prove useful to practitioners—namely, apprenticeship. The apprenticeship notion emerges from the literature of anthropology. In particular, much of the current discussion about classroom practice is an extension of Lave's (1977) study of the apprenticeship of tailors in Liberia, in which novices worked with master tailors to learn the skills, habits, and dispositions of these masters toward their work and to develop their own knowledge and skills by working on "real" tailoring tasks. Applying this view to school mathematics, Lave, Smith, and Butler (1988) have recently argued that the purpose of the activity of school mathematics might be seen not as the communication of decontextualized and abstract skills and concepts but rather as the development of a richly textured knowledge base in which knowledge is situated in important intellectual tasks. The apprenticeship view also suggests that a major purpose of school mathematics is to develop in students the habits of thinking and the points of view of professionals in the field; that is, the goal is learning to think mathematically.

Obviously, this view needs to be carefully worked out for students at various grade levels, but an apprenticeship perspective on mathematics education offers a vision of mathematics classrooms as places where students, under the careful tutelage of their mathematics teacher, engage in *doing* mathematics rather than having it done to them. Therein lies the connection to another research perspective that is beginning to emerge as an important one for practitioners to consider: socially constructed and socially shared knowledge. Resnick (1988) has argued that it may be productive to consider mathematics not as a subject that is well-structured and tightly organized but rather as a knowledge domain that is subject to interpretation and meaning construction. She advocates the use of social settings, such as socially shared problem solving, as arenas of argument and debate leading to the development of shared mathematical meaning. Moreover, she argues that this view of mathematics teaching should be applicable to all aspects of mathematics instruction, not simply to problem-solving activity. Resnick's analysis suggests a view of mathematics classrooms that makes close contact not only with the apprenticeship view of mathematics classrooms but also with research on the social and cultural contexts of mathematics education (e.g., Bishop 1988). Moreover, this perspective generally appears to be quite compatible with many of the themes emphasized in *Curriculum and Evaluation Standards for School Mathematics* (NCTM 1989), especially those dealing with the importance of social interaction, the communication of mathematical ideas, and the importance of providing students with experiences in problem posing, conjecturing, and other generative cognitive activities.

This view of mathematics classrooms as places in which classroom activity is directed not solely toward the acquisition of the content of mathematics in the form of concepts and procedures but also toward the situated, collaborative practice of mathematical thinking is not completely new to mathematics education. Similar views were previously found in the writings of some mathematicians and matematics educators (e.g., Fawcett 1938; Polya 1965). Until recently, however, this view has not found a suitable characterization in the research community. It is likely that the emergence of this view, compatible with the vision painted by the NCTM's *Curriculum and Evaluation Standards,* in the thinking and writing of many researchers will make their new work even more useful for practice.

RELATING RESEARCH AND PRACTICE: A BIDIRECTIONAL RELATIONSHIP

In this article, we have considered aspects of research that have potential relevance for the improvement of educational practice. It is important, however, to note that the relationship between research and practice is bidirectional. Practitioners will want to consider the multiple ways in which research can positively influence their instructional thinking and actions, but it is equally important for researchers to consider and respond to problems and issues that are raised when practitioners apply research findings, methods, or theoretical perspectives. Moreover, it is important for researchers and practitioners to become collaborators in investigating issues of practical importance for the improvement of the teaching and learning of mathematics. If the recommendations made in NCTM's *Curriculum and Evaluation Standards* are to become realities, then researchers and practitioners will need to work closely together in the transformative process. Although the goals of the two groups are often not identical, our collective understanding of mathematics learning and teaching will benefit from partnerships of researchers and practitioners theorizing, planning, conducting, and interpreting research that is both educationally relevant and scientifically respectable.

REFERENCES

Bishop, Alan J. *Mathematical Enculturation.* Dordrecht, Netherlands: Kluwer, 1988.

Brown, Stephen I., and Marion I. Walter. *The Art of Problem Posing.* Hillsdale, N.J.: Lawrence Erlbaum Associates, 1983.

Carpenter, Thomas P., James M. Moser, and Thomas Romberg, eds. *Addition and Subtraction: A Cognitive Perspective.* Hillsdale, N.J.: Lawrence Erlbaum Associates, 1982.

Charles, Randall I., and Edward A. Silver, eds. *The Teaching and Assessing of Mathematical Problem Solving*. Research Agenda for Mathematics Education, vol. 3. Reston, Va.: National Council of Teachers of Mathematics, 1988.

Chi, Michelene T. H., Miriam Bassok, Matthew W. Lewis, Peter Reimann, and Robert Glaser. "Self-Explanations: How Students Study and Use Examples in Learning to Solve Problems." *Cognitive Science* 13 (1989): 145–82.

Fawcett, Harold. *The Nature of Proof*. Thirteenth Yearbook of the National Council of Teachers of Mathematics. New York: Columbia University, Teachers College, Bureau of Publications, 1938.

Garofalo, Joe, and Frank K. Lester, Jr. "Metacognition, Cognitive Monitoring, and Mathematical Performance." *Journal for Research in Mathematics Education* 16 (May 1985): 163–76.

Grouws, Douglas A., Thomas J. Cooney, and Douglas Jones, eds. *Perspectives on Research on Effective Mathematics Teaching*. Research Agenda for Mathematics Education, vol. 1. Reston, Va.: National Council of Teachers of Mathematics, 1988.

Hiebert, James, and Merlyn J. Behr, eds. *Number Concepts and Operations in the Middle Grades*. Research Agenda for Mathematics Education, vol. 2. Reston, Va.: National Council of Teachers of Mathematics, 1988.

Kilpatrick, Jeremy. "Problem Formulating: Where Do Good Problems Come From?" In *Cognitive Science and Mathematics Education,* edited by Alan H. Schoenfeld, pp. 123–48. Hillsdale, N.J.: Lawrence Erlbaum Associates, 1987.

Lave, Jean. "Tailor-made Experiments and Evaluating the Intellectual Consequences of Apprenticeship Training." *Quarterly Newsletter of the Institute for Comparative Human Development 1*(2) (1977): 1–3.

Lave, Jean, Steven Smith, and Michael Butler. "Problem Solving as Everyday Practice." In *The Teaching and Assessing of Mathematical Problem Solving,* edited by Randall I. Charles and Edward A. Silver, pp. 61–81. Reston, Va.: National Council of Teachers of Mathematics, 1988.

National Council of Teachers of Mathematics. *Curriculum and Evaluation Standards for School Mathematics*. Reston, Va.: The Council, 1989.

Owen, Elizabeth, and John Sweller. "What Do Students Learn While Solving Mathematics Problems?" *Journal of Educational Psychology* 77 (1985): 271–84.

Polya, George. *Mathematical Discovery*. Vol. 2. New York: John Wiley & Sons, 1965.

Resnick, Lauren B. "Treating Mathematics as an Ill-structured Discipline." In *The Teaching and Assessing of Mathematical Problem Solving,* edited by Randall I. Charles and Edward A. Silver, pp. 32–60. Reston, Va.: National Council of Teachers of Mathematics, 1988.

Schoenfeld, Alan H. *Mathematical Problem Solving*. Orlando, Fla.: Academic Press, 1985.

———. "What's All the Fuss about Metacognition?" In *Cognitive Science and Mathematics Education,* edited by Alan H. Schoenfeld, pp. 189–216. Hillsdale, N.J.: Lawrence Erlbaum Associates, 1987.

Schoenfeld, Alan H., and Douglas Herrmann. "Problem Perception and Knowledge Structure in Expert and Novice Mathematical Solvers." *Journal of Experimental Psychology: Learning, Memory, and Cognition* 8 (1982): 484–94.

Silver, Edward A. "Student Perceptions of Relatedness among Mathematical Verbal Problems." *Journal for Research in Mathematics Education* 10 (May 1979): 195–210.

Silver, Edward A., ed. *Teaching and Learning Mathematical Problem Solving: Multiple Research Perspectives*. Hillsdale, N.J.: Lawrence Erlbaum Associates, 1985.

Silver, Edward A., and Verna M. Adams. "Using Open-ended Problems." *Arithmetic Teacher* 34 (May 1987): 34–35.

Silver, Edward A., Nicholas Branca, and Verna M. Adams. "Metacognition: The Missing Link in Problem Solving?" In *Proceedings of the Fourth International Conference for the Psychology of Mathematics Education,* edited by R. Karplus, pp. 213–30. Berkeley, Calif.: University of California, 1980.

Silver, Edward A., and J. Philip Smith. "Think of a Related Problem." In *Problem Solving in School Mathematics,* 1980 Yearbook of the National Council of Teachers of Mathematics, edited by Stephen Krulik, pp. 146–56. Reston, Va.: The Council, 1980.

Smith, Margaret S., and Edward A. Silver. "Cancelling Cancellation: The Role of Worked-out Examples in Unlearning a Procedural Error." In *Proceedings of the 11th Annual Meeting of the North American Chapter of the International Group for the Psychology of Mathematics Education,* edited by Carolyn A. Maher, Gerald A. Goldin, and Robert B. Davis, pp. 40–46. New Brunswick, N.J.: Center for Math, Science and Computer Education, Rutgers University, 1989.

Sweller, John, and Graham A. Cooper. "The Use of Worked Examples as a Substitute for Problem Solving in Learning Algebra." *Cognition and Instruction* 2 (1985): 59–89.

Sweller, John, and Maita Levine. "Effects of Goal Specificity on Means-Ends Analysis and Learning." *Journal of Experimental Psychology: Learning, Memory, and Cognition* 8 (1982): 463–74.

Sweller, John, Robert F. Mawer, and Mark R. Ward. "Development of Expertise in Mathematical Problem Solving." *Journal of Experimental Psychology: General* 112 (1983): 634–56.

Wagner, Sigrid, and Carolyn Kieran. *Research Issues in the Learning and Teaching of Algebra.* Research Agenda for Mathematics Education, vol. 4. Reston, Va.: National Council of Teachers of Mathematics, 1989.

Weaver, J. Fred. "Big Dividends from Little Interviews." *Arithmetic Teacher* 2 (April 1955): 40–47.

Zhu, Xinming, and H. A. Simon. "Learning Mathematics from Examples and by Doing." *Cognition and Instruction* 4 (1987): 137–66.

2

The Importance of Social Interaction in Children's Construction of Mathematical Knowledge

Erna Yackel
Paul Cobb
Terry Wood
Grayson Wheatley
Graceann Merkel

When children learn mathematics in school, they do so in a classroom where certain standards of conduct are established either explicitly or implicitly. These standards, or norms, influence the way children interact with the teacher and with each other, which in turn influences both what mathematics the children learn and how they learn it. In this article we shall discuss an instructional approach in which the role of social interaction in children's learning of mathematics is given explicit attention. The approach reflects the view that learning mathematics is an active, problem-solving process. When children are given opportunities to talk about their mathematical understandings, problems of genuine communication arise. These problems, as well as the mathematical tasks themselves, constitute occasions for learning mathematics. The approach we shall discuss is consistent with current calls for reform in mathematics education (National Research Council 1989), with the principles that guided the development of the National Council of Teachers of Mathematics *Curriculum and Evaluation Standards for School Mathematics* (NCTM 1989), and with principles of constructivist

---

The project discussed in this article is supported by the National Science Foundation under Grant No. MDR-847-0400. All opinions expressed are, of course, solely those of the authors.

learning theory in general (Confrey 1987; von Glasersfeld 1984). A detailed account of the problem-centered aspect of the instructional approach can be found in Cobb, Wood, and Yackel (forthcoming). Quantitative results of the project for second graders on which this chapter is based are presented in Nicholls et al. (in press) and Nicholls et al. (forthcoming).

Our focus in this article will be on three aspects of the instructional approach: First, children's construction of their own nonstandard methods; second, mathematical learning as problem-solving activity; and finally, the role of social interaction in learning mathematics. The first two sections provide a background for the discussion of the third.

CHILDREN CONSTRUCT THEIR OWN MATHEMATICS

When children are presented with tasks and encouraged to solve them in ways that make sense to them rather than follow procedures that have been presented by the teacher, they develop a variety of solution methods. Early in the school year, children offered the following solution methods for solving 9 + 11 = __:

Brenda: 9 and 9 is 18, plus 2 is 20.
Adam: 7 and 7 is 14, so 8 and 8 is 16. 9 and 9 would be 18 so 9 + 11 must equal 20.
Chris: 11 and 11 equals 22. 10 and 11 equals 21. 9 and 11 equals 20.
Jane: 11 and 9 more—12, 13 , . . . , 18, 19, 20.

As these examples illustrate, in a conducive setting children use what they already know to develop personally meaningful solutions. Their solutions reflect differences in the current knowledge they bring to the task.

We contend that not only are children capable of developing their own methods for completing school mathematics tasks but that each child has to construct his or her own mathematical knowledge. That is, in our view, mathematical knowledge cannot be *given* to children. Rather, they develop mathematical concepts as they engage in mathematical activity, including trying to make sense of methods and explanations they see or hear from others. The implications of this view for instruction are that children should be provided with activities that are likely to give rise to genuine mathematical problems. These problems give them the opportunity to reflect and reorganize their current ways of thinking. Accordingly, we developed instructional activities designed to foster children's construction of relatively sophisticated concepts and procedures, such as place value and computational algorithms.

The following four efficient solutions to 39 + 53 = __ illustrate the conceptually based nature of children's alternative algorithms:

Anna: 50 plus 30—80, then 9 plus 1 more would be 90, plus 2 more would be 92.

Joel: You have 53, 10 more is 63, plus 10 more—73, plus 10 more 83, plus 9 . . . 92.

Jenny: See, 39 and 50 more is 89, then add 3 makes 92.

Eric: 30 plus 50 is 80 and 9 plus 3 is 12. Put all those together and I came up with 92.

The rich variety in the algorithms that children develop when they engage in meaningful thinking contrasts with the blind rule-following we so often see when children are trained to use one particular algorithm. It is the difference between "What was I told I am supposed to do?" and "How can I figure this out?" An important consequence of our approach is that some types of errors that commonly accompany use of the traditionally taught algorithms and that frequently result in unreasonable answers (e.g., 39 + 53 = 812) rarely occur because children are using methods that are based on their understandings. Our experience is that children's understanding of place value develops together with their construction of increasingly efficient algorithms. We concur with Brownell (1956) that conceptual understanding and computational proficiency, or meaning and practice as he described it, should not be seen as separate instructional goals (Cobb, Yackel, and Wood 1988).

LEARNING AS A PROBLEM-SOLVING ACTIVITY

A chief feature of an instructional approach that is based on the constructivist view of learning is that instructional activities should give rise to problems for students to resolve (Cobb, Wood, and Yackel forthcoming; Thompson 1985). However, the situations that children find problematic differ owing to wide differences in their knowledge, experiences, and goals. At first glance, this may appear to be a limitation, since we cannot guarantee that all children will think about a task in the same way. In fact, it becomes an advantage in that it is a means of individualization. Children at differing conceptual levels not only use different solution methods but interpret tasks in different ways. In essence, each child attempts to solve problems that make sense given his or her level of understanding and conceptual development. It is in this sense that we say that teachers cannot give problems to students ready made. Teachers can give instructional activities. The problems children solve will differ from child to child. The following example of a money story problem from the classroom illustrates this point:

> Bruce has this much money. He buys a pencil for 7 cents and a candy bar for 35 cents. How much money is left?

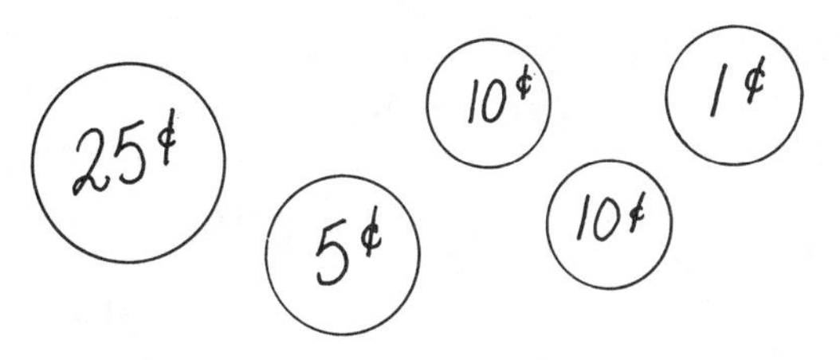

Lois's solution is to take away the quarter and one dime for the candy bar (she crossed these off on the diagram) and then to take away the nickel, the penny, and "one away from the 10" (the remaining dime) for the pencil to give an answer of 9 cents. Adam, though, solved the problem as follows:

> I added them all up and I got 51 and then I minused the 35. [He then went on to subtract 7.]

Adam conceptualized the task as a symbolic addition and subtraction problem. Lois, in contrast, relied on the real-world setting and used the value of coins. Like Adam, she engaged in mental computation, but her problem was to take away coins that equaled the value of the purchased items and so find the value of the remaining coins. Both children engaged in meaningful mathematical activity on the same task despite differences in their conceptual knowledge.

Solving problems often involves more than figuring out a way to complete the instructional activities. It can also include accounting for surprising outcomes, such as when two alternative methods lead to the same result, justifying a solution method, or explaining why an apparently erroneous method leads to a contradiction. When classroom instruction is organized so that children work cooperatively on the instructional activities, situations such as these occur frequently. In the next section we shall discuss the role of social interaction in learning.

LEARNING THROUGH SOCIAL INTERACTION

We shall first outline the instructional approach and then illustrate typical teacher-student and student-student interactions.

The Instructional Approach

The instructional activities are of two general types, teacher-orchestrated whole-class activities and small-group activities. In a typical fifty-minute class period, the first half is devoted to small-group problem solving and the second half to a class discussion in which children explain how they solved the activities. The introduction of the activities is limited to ensuring that the children understand their intent and are familiar with any symbols used. It does not include any explanation or demonstration by the teacher of how to solve the activities. During small-group work the children, working in pairs, are expected to cooperatively develop solutions to the activities. Meanwhile, the teacher circulates among the groups observing and intervening in their problem-solving efforts.

In the subsequent class discussion the children explain how they solved the activities. The teacher helps children clarify their explanations, assists them as they verbalize their thinking, and actively encourages them to present alternative solutions. The teacher does not tell children if their answers are correct or incorrect but encourages all of them to reflect on the solutions presented and agree or disagree. When children disagree, the class works as a whole to resolve the disagreement and reach consensus. Some problems remain unresolved for several days during which children can often be seen having extensive discussions about them during recess or lunch time. At the conclusion of the class discussion, the teacher collects the children's activity pages, dates them, and places them in the children's individual folders, which are sent home periodically.

The instructional activities presented to the children follow a variety of formats but are all designed to facilitate the occurrence of mathematical problems for the children to resolve. Detailed models of children's early number learning (Steffe, von Glasersfeld, Richards, and Cobb 1983; Steffe, Cobb, and von Glasersfeld 1988) were used in the development of the activities.

Teacher-Student Interaction

The teacher's attitude is crucial to the development of a problem-solving atmosphere in the classroom. In order for children to share their mathematical thinking, they must actively attempt to communicate with each other and with the teacher. Successful communication requires the negotiation of meanings (Bishop 1985) and "depends on all members of the class expressing genuine respect and support for one another's ideas" (NCTM 1989, p. 29). With respect to the approach advocated here, this means that every time a child offers a comment in the class discussion, the teacher assumes that the mathematical activity the child is attempting to describe is meaningful to that child. It becomes the responsibility of the teacher to try to figure out what the child means and, if necessary, to assist the child in verbalizing this meaning. The importance of the teacher's attempts to ascribe meaning is illustrated with the following example.

In this example, taken from a lesson in early December, the class was working on a whole-class activity in which the teacher asked the children to try to figure out the answer to $9 + 9 + 9$ without counting. The first child to offer a solution added 10 plus 10 plus 10 and then subtracted three ones to get 27. The next solution was offered by Mike:

> I changed one 9 to a 10 and one 9 into a 17 and then I took away one 9 and then I came up with 27.

Instead of treating this as a confused and muddled manipulation of numbers, the teacher assumed that Mike was attempting to communicate meaningful thinking. First she signaled this assumption with her comments, and then she proceeded to try to figure out what Mike might have been trying to say:

> *Teacher:* Did you hear that, Jennie? Run that—okay, now listen. I may need your help. [To Mike] I want you to say it again. You said you took one 9 and turned it into a 10.
>
> *Mike:* And I took nine—I took seve . . .—and changed the second 9 into a 17.
>
> *Teacher:* Changed that [pointing to the second 9] into a 17.
>
> *Mike:* And I took away that last 9 into an equals and then I came up with 27.

Here again the teacher might assume that Mike does not know what he is talking about. However, she continues to assume that what he did made sense to him, and she attempts to help him develop an explanation that might make sense to the other children.

> *Teacher:* [to the class] Do you see how he did it?
>
> *Adam:* He took away—it away from the last 9 and put it to that [the first 9]. Added that 8 to the 9 and got 17.
>
> *Teacher:* Let's take a look at this part of it [pointing to the last two 9s]. What's 9 + 9?
>
> *Students:* [in unison] 18.
>
> *Teacher:* 18. Okay. He knew that 9 and 9 made 18. Okay. He knew that. Now, he took 1 away from 18 and what does that give us?
>
> *Students:* [in unison] 17.
>
> *Teacher:* 17, and he took the 1 that he had over here (taken away from the 17) and he added it to this 9 (the first 9) and that made 10. That was a way to figure it out.

Although it is impossible for us to tell if the solution the teacher developed is what Mike had in mind, her pursuing the discussion with him had several positive effects. First, the teacher made it clear to the children that she assumes their solutions make sense. Second, Mike and the other children in the class were reassured that the teacher would help them as they try to verbalize their solution attempts. Finally, the entire class had the benefit of thinking through another solution method.

When a child gives an incorrect answer, it is especially important for the teacher to assume that the child was engaged in meaningful activity. Thus, it is possible that the child will reflect on his or her solution attempt and evaluate it. When a child gives an incorrect answer in this project, it is not uncommon that, without assistance or prompting, he or she finds the error in the course of explaining the solution. As one child said in such a situation, "Now I disagree with my own answer." By allowing a child to proceed with an explanation even when the answer is wrong, the teacher fosters a belief that the teacher is not the sole authority in the classroom to whom children have to appeal to find out if their answers are right or wrong. Children are able to make such decisions for themselves. Mathematical authority does not reside solely with the teacher, but with the teacher and the children as an intellectual community.

Not only does the teacher assume that what the children are doing and saying makes sense to them, she also expects the children to make that assumption about each other. She strives to oblige them to make sense of one another's solution attempts. This applies both in the small-group work and in the total-class discussions. Adam's remarks in the example above indicate that as early as second grade, children can begin to accept this obligation. Their role in class discussions is to listen and reflect on what is being said and try to make sense of it in terms of their own cognitive framework. In the example Adam apparently thinks he understands Mike's solution method. By entering into the discussion, Adam derives some benefits for himself as he clarifies his own thinking. Here we concur with Barnes and Todd (1977) in assuming "that speech functions as a means by which people construct and reconstruct views of the world about them, often jointly, when the speech is a means of communication with other people" (p. 1). At other times, children might interject comments such as, "I don't understand what she is trying to say" or "Oh, I see how she did it," giving evidence that the listeners are trying to understand what the speaker is saying.

The teacher's posture of trying to figure out what sense the children are trying to make of the activities also applies in small-group problem solving. In fact, it is one of the most important principles in guiding the teacher's intervention in small groups. When approaching a small group at work, the teacher's first responsibility is to try and figure out how the children are thinking about their task. Only then does the teacher engage in a discussion with them. Her or his role in such interventions, as in the class discussion, is not to tell the children if they are right or wrong or to lead them to a correct solution, but rather to assist them as they attempt to develop mathematical meanings. This means that the teacher must judge what type of assistance, if any, is appropriate. It may be to encourage the children to work cooperatively or to listen to one another's explanations. It may be to ask the children provocative questions or enter into a Socratic dialogue with them. It may be to assist one of them to explain his or her thinking to another, or it may be to facilitate a dialogue (Wood and Yackel forthcoming).

Student-Student Interaction

Children engage in two types of problem solving as they work together in small groups to complete the instructional activities. On the one hand, they attempt to solve their mathematical problems; on the other hand, they have to solve the problem of working productively together. The obligations the teacher attempts to place on the children as they work in small groups are (1) that they should cooperate to solve the problems and (2) that they should reach a consensus. These two obligations mean that children should explain their thinking to one another, that they try to understand one another's thinking, that they assume one another's solution attempts make sense, and that they persist in trying to figure things out for themselves.

The interactions that take place once the problem of social cooperation has been temporarily resolved give rise to opportunities for learning that do not typically occur in traditional classroom settings (Yackel, Cobb, and Wood in press), including opportunities for children to verbalize their thinking, explain or justify their solutions, and ask for clarifications. Attempts to resolve conflicts lead to opportunities for children to reconceptualize a problem and to extend their conceptual framework to incorporate alternative solution methods.

The following example illustrates how two children extended their conceptual frameworks as they attempted to resolve a conflict as they solved $39 + 19 = \_\_$.

Craig: [Uses a hundreds board and starts counting at 40 on the board.] 40, 41, . . . , 57, 58, 59. [While counting, he does not visibly keep track of his counting acts.]

Karen: 39, 49. That's ten [pointing to the hundreds board. She continues counting on her fingers starting at 50. She puts up one more finger with each number word utterance and stops when nine fingers are up.] 50, 51, . . . , 57, 58.

In the process of trying to resolve the disagreement between their answers, each child repeats his or her solution several times. Finally Karen figures out the possible source of Craig's difficulty.

Karen: You're not even counting [meaning keeping track of your counting]. Come here. I'll explain how I got my number. See, you have 39 and you plus 10 more and that's 49. 50, 51, . . . , 58. [This time Karen is counting on the hundreds board by pointing to numerals on it with her pencil. Simultaneously, she uses fingers of both hands to keep track of her counting acts. She stops when nine fingers are up.]

Here Karen reconceptualized her own solution in light of Craig's method and adapted her explanation to use the hundreds board as Craig had done. In doing so, she had to draw on her own understanding to develop a framework within which to give an explanation that might make sense to Craig. In a lengthy ensuing dialogue Craig eventually extended his initial conceptualization as he made sense of Karen's explanations, and the conflict was resolved.

CONCLUSION

We have discussed an instructional approach that is based on the view that mathematics is a creative human activity and that social interaction in the classroom plays a crucial role as children learn mathematics. Both the interaction between teacher and child and the interactions among children influence what is learned and how it is learned. The teacher plays a crucial role by guiding the development of what Silver (1985) called a problem-solving atmosphere, an environment in which children feel free to talk about their mathematics. The teacher also is instrumental in structuring "a pervasive norm in the classroom that helping one's peers to learn is not a marginal activity, but is a central element of students' roles" (Slavin 1985, p. 16). Once this norm is established, opportunities for learning, not present in traditional classrooms, arise as children collaborate to solve problems.

As a final point, we note that children learn a lot more than mathematics in this type—or any type—of classroom setting. They develop beliefs about mathematics and about their own and the teacher's role. In addition, a sense of what is valued is developed along with attitudes and forms of motivation. The approach described above is designed to foster the beliefs that persisting and figuring out a personally challenging problem are valued more than a page of correct answers, that meaningful activity is valued over what one child called "mixing up a bunch of numbers," and that cooperation and negotiation are valued over competition and conflict (Cobb, Yackel, and Wood 1989). Above all else, the approach of encouraging students to talk about their solution methods without evaluating them for correctness is characterized by the development of a mutual trust between the teacher and the students. The teacher trusts the students to persist in attempting to resolve their mathematical problems and consequently feels free to call on them to describe their thinking. The students trust the teacher to respect their efforts and consequently enter into discussions in which they explain how they actually understood and attempted to solve their mathematical problems.

REFERENCES

Barnes, Douglas, and Frankie Todd. *Communication and Learning in Small Groups*. London: Routledge & Kegan Paul, 1977.

Bishop, Alan. "The Social Construction of Meaning—a Significant Development for Mathematics Education?" *For the Learning of Mathematics* 5 (1) (1985): 24–28.

Brownell, William A. "Meaning and Skill: Maintaining the Balance." *Arithmetic Teacher* 3 (October 1956): 129–36.

Cobb, Paul, Terry Wood, and Erna Yackel. "Learning through Problem Solving: A Constructivist Approach to Second-Grade Mathematics." In *Constructivism in Mathematics Education*, edited by Ernst von Glasersfeld. Dordrecht, Netherlands: D. Reidel, forthcoming.

Cobb, Paul, Erna Yackel, and Terry Wood. "Curriculum and Teacher Development: Psychological and Anthropological Perspectives." In *Integrating Research on Teaching and Learning Mathematics*, edited by Elizabeth Fennema, Thomas Carpenter, and Sue Lamon. Madison: Wisconsin Center for Education Research, 1988.

________. "Young Children's Emotional Acts While Doing Mathematical Problem Solving." In *Affect and Mathematical Problem Solving: A New Perspective*, edited by Douglas B. McLeod and Verna M. Adams. New York: Springer-Verlag, 1989.

Confrey, Jere. "The Current State of Constructivist Thought in Mathematics Education." Paper presented at the annual meeting of the International Group for Psychology of Mathematics Education, Montreal, July 1987.

National Council of Teachers of Mathematics. *Curriculum and Evaluation Standards for School Mathematics*. Reston, Va.: NCTM, 1989.

National Research Council. *Everybody Counts: A Report to the Nation on the Future of Mathematics Education*. Washington, D.C.: National Academy Press, 1989.

Nicholls, John G., Paul Cobb, Terry Wood, Erna Yackel, and Michael Patashnick. "Dimensions of Success in Mathematics: Individual and Classroom Differences." *Journal for Research in Mathematics Education*, in press.

Nicholls, John G., Paul Cobb, Terry Wood, Erna Yackel, and Grayson Wheatley. "Assessing Young Children's Mathematical Learning." In *Assessing Higher Order Thinking in Mathematics*, edited by Gerald Kulm. Washington D.C.: American Association for the Advancement of Science, forthcoming.

Silver, Edward A. "Research on Teaching Mathematical Problem Solving: Some Underrepresented Themes and Needed Directions." In *Teaching and Learning Mathematical Problem Solving: Multiple Research Perspectives*, edited by Edward A. Silver, pp. 247–66. Hillsdale, N.J.: Lawrence Erlbaum Associates, 1985.

Slavin, Robert. "An Introduction to Cooperative Learning Research." In *Learning to Cooperate, Cooperating to Learn*, edited by Robert Slavin, Shlomo Sharan, Spencer Kagan, Rachel Hertz-Lazarowitz, Clark Webb, and Richard Schmuck, pp. 5–15. New York: Plenum Press, 1985.

Steffe, Leslie P., Paul Cobb, and Ernst von Glasersfeld. *Young Children's Construction of Arithmetical Meanings and Strategies*. New York: Springer-Verlag, 1988.

Steffe, Leslie P., Ernst von Glasersfeld, John Richards, and Paul Cobb. *Children's Counting Types: Philosophy, Theory, and Application*. New York: Praeger Scientific, 1983.

Thompson, Patrick. "Experience, Problem Solving, and Learning Mathematics: Considerations in Developing Mathematics Curricula." In *Teaching and Learning Mathematical Problem Solving: Multiple Research Perspectives*, edited by Edward A. Silver, pp. 189–236. Hillsdale, N.J.: Lawrence Erlbaum Associates, 1985.

von Glasersfeld, Ernst. "An Introduction to Radical Constructivism." In *The Invented Reality*, edited by Paul Watzlawick, pp. 17–40. New York: W. W. Norton & Co., 1984.

Wood, Terry, and Erna Yackel. "The Development of Collaborative Dialogue within Small-Group Interactions." In *Transforming Early Childhood Mathematics*, edited by Leslie P. Steffe and Terry Wood. Hillsdale, N.J.: Lawrence Erlbaum Associates, forthcoming.

Yackel, Erna, Paul Cobb, and Terry Wood. "Small-Group Interactions as a Source of Learning Opportunities in Second-Grade Mathematics." In *Cooperative Learning in Mathematics*, *Journal for Research in Mathematics Education Monograph*, forthcoming.

3

Constructivism and Beginning Arithmetic (K–2)

Constance Kamii

DO CHILDREN acquire number concepts by having them taught? Do they learn arithmetic by internalizing rules (algorithms)? Piaget's answer to these questions is an unequivocal no. His theory, called constructivism, demonstrates that human beings acquire knowledge by building it from the inside instead of internalizing it directly from the environment. The best way to clarify this statement is with examples of children's reactions to a Piagetian task.

CONSTRUCTIVISM

A Piagetian Task

In this task, which was devised by Bärbel Inhelder and Jean Piaget (1963), two identical glasses and about fifty wooden beads (or chips, beans, etc.) are used. The child is given one of the glasses, and the teacher takes the other glass. The teacher then asks the child to drop a bead into his or her glass each time the teacher drops one into the other glass. After about five beads have thus been dropped into each glass with one-to-one correspondence, the adult says, "Let's stop now, and you watch what I am going to do." The teacher then drops one bead into the glass and says to the child, "Let's get going again." The adult and the child drop about five more beads into each glass with one-to-one correspondence, until the adult says, "Let's stop." The following is what has happened so far:

Adult: 1 + 1 + 1 + 1 + 1 + 1 + 1 + 1 + 1 + 1 + 1
Child: 1 + 1 + 1 + 1 + 1 + 1 + 1 + 1 + 1 + 1

The adult now asks, "Do we have the same amount, or do *you* have more, or do *I* have more?"

Four-year-olds usually reply that the two glasses have the same amount. When we go on to ask, "How do you know that we have the same amount?" the children explain, "Because I can see that we both have the same (amount)." (Some four-year-olds, however, reply that *they* have more, and when we ask them how they know that they have more, their explanation consists of only one word: "Because.")

The adult goes on to ask, "Do you remember how we dropped the beads?" and four-year-olds usually give all the empirical facts correctly, including the fact that only the adult put one bead into the glass at one point. In other words, four-year-olds remember all the empirical facts correctly and base their judgment of equality on the empirical appearance of the two quantities.

By age five or six (in kindergarten), however, most children deduce logically that the teacher has one more. When we ask these children how they know that the adult has one more, they invoke exactly the same empirical facts as the four-year-olds.

If a child says that the adult's glass has one more bead, the teacher goes on to pose the next question: "If we continued to drop beads all day (or all night) in the same way (with one-to-one correspondence), do you think we will have the same number at the end, or will *you* have more, or will *I* have more?" Five- and six-year-olds divide themselves into two groups at this point. The more advanced group answers in the way that adults would, that is, that there will *always* be one more in the teacher's glass. The other group makes empirical statements such as "I don't know because we haven't done it yet" or "We don't have enough beads to keep going all day."

Children's responses in this task illustrate their construction of number concepts from the inside. No one teaches five- and six-year-olds to give correct answers to these questions. Yet children all over the world become able to give correct answers by constructing numerical relationships through their own natural ability to think. This construction from within can best be explained by reviewing the distinction Piaget made among three kinds of knowledge—physical knowledge, logico-mathematical knowledge, and social (conventional) knowledge.

Physical, Logico-Mathematical, and Social Knowledge

Physical knowledge is knowledge of objects in external reality. The color and weight of a bead are examples of physical properties that are *in* objects in external reality and that can be known empirically by observation.

Logico-mathematical knowledge, by contrast, consists of *relationships* created by each individual. For instance, when we are presented with a red bead and a blue one and think that they are *different,* this difference is an example of logico-mathematical knowledge. The beads are observable, but the *difference* between them is not. The difference exists neither *in* the red bead nor *in* the blue one, and if a person did not put the objects into this relationship, the difference would not exist for him or her. Other examples of relationships the individual can create between the same beads are "similar" and "two."

Physical knowledge is thus empirical knowledge that has its source partly in objects. Logico-mathematical knowledge is not empirical knowledge, since its source is in each individual's head.

The ultimate sources of social knowledge are conventions worked out by people. Examples of social knowledge are the fact that Christmas comes on 25 December and that a tree is called "tree." (For a more complete discussion of the three kinds of knowledge, the reader is referred to Kamii [1982, 1985, 1989].)

The distinction among the three kinds of knowledge makes it possible to understand why most four-year-olds in the task described earlier say that the two glasses have the same amount. When children have not yet constructed the logico-mathematical relationships of number in their heads, all they can get from the experience is physical, empirical knowledge. This is why four-year-olds can remember the empirical facts of dropping all the beads except one with one-to-one correspondence. This one-to-one correspondence, however, is only empirical, and four-year-olds judge the quantity of beads also empirically. This is why they say that the two glasses have the same amount and explain, "I can *see* they have the same amount."

By age five or six, however, most children have constructed the logico-mathematical knowledge of number and can deduce from the same empirical facts that the teacher has one more bead. However, this structure of number takes many years to construct, and the child who has number concepts up to ten or fifteen does not necessarily have concepts of fifty, a hundred, or more.

The Child's Construction of Number Concepts

According to Piaget, the child constructs number concepts by synthesizing two kinds of relationships—order and hierarchical inclusion. Let us examine order first and then hierarchical inclusion.

All teachers of young children have noticed four-year-olds' tendency to "count" objects by skipping some and counting others more than once. When given eight objects, for example, a four-year-old who can recite the number names to twelve may claim that there are twelve by counting the same objects several times and skipping others. This behavior indicates that the child does not feel the logical necessity of putting the objects into an ordered relationship to make sure none are skipped or counted more than once. The only way we can be sure of not overlooking any or counting some more than once is by putting them into some relationship of order, like that shown in figure 3.1. The child does not, however, have to put the objects in a linear order; what is important is that he or she order them *mentally* in some way.

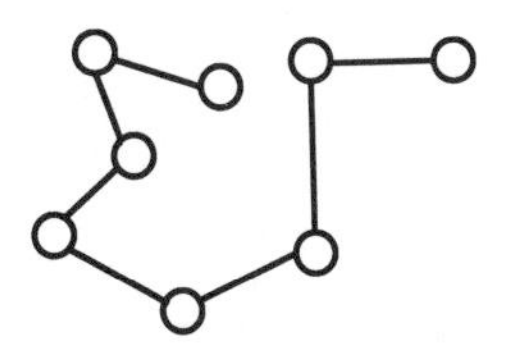

Fig. 3.1. An example of a mental ordering of objects

If ordering were children's only mental action on the objects, they would not be able to quantify the collection numerically. For example, after counting eight objects arranged in linear order, as shown in figure 3.2(a), four-year-olds usually say that there are eight. If we then ask them to show us eight, they sometimes point to the last one (the eighth object). This behavior indicates that for these children, the words "one, two, three . . ." are names for individual elements in a series, like "Monday, Tuesday, Wednesday. . . ."

To quantify the collection of objects, children have to put them into a relationship of hierarchical inclusion. This relationship, shown in figure 3.2(b), means that children have to *mentally* include "one" in "two," "two" in "three," "three" in "four," and so on. When presented with eight objects, they can quantify the collection numerically only if they can put all the objects into a single relationship, thus synthesizing order and hierarchical inclusion.

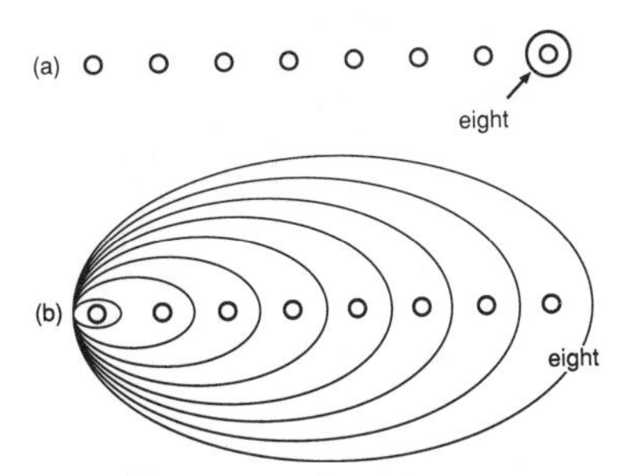

Fig. 3.2. Hierarchical inclusion (b) and its absence (a)

I hope the reader can see that contrary to traditional assumptions, children do not acquire number concepts by abstracting them from sets of objects (as if number concepts were physical knowledge). Rather, children become able to think about the objects as "eight" only when they can impose their logico-mathematical knowledge, that is, self-created relationships, on the set.

It must be obvious from the preceding discussion that a theory stating that children construct logico-mathematical knowledge from the inside has far-reaching implications for changes in the way arithmetic is taught. Let us now turn to the implications of constructivism for teaching arithmetic in grades K–2.

IMPLICATIONS OF CONSTRUCTIVISM FOR TEACHING BEGINNING ARITHMETIC

The fact that children construct their own logico-mathematical knowledge does not imply that the teacher's role is to sit back and do nothing. On the contrary, the teacher's role becomes more indirect and difficult than in traditional instruction. Two of the most important implications of constructivism for teaching beginning arithmetic follow.

1. *We must focus our goals and objectives on children's thinking rather than on their writing correct answers.*

The implicit goal of instruction with textbooks, workbooks, and worksheets is to get children to *write* correct answers. Already at the beginning of first grade, children are taught how to write correct answers to exercises such as the one shown in figure 3.3(a). In second grade, when double-column addition with regrouping is introduced, the goal of becoming able to write correct answers involves rules (called *algorithms*) that are introduced with frames like the one in figure 3.3(b). This frame reduces the addition of two-digit numbers to the mechanical writing of numerals by following rules children do not understand. The very need to present such a frame proves that children do not understand regrouping and place value.

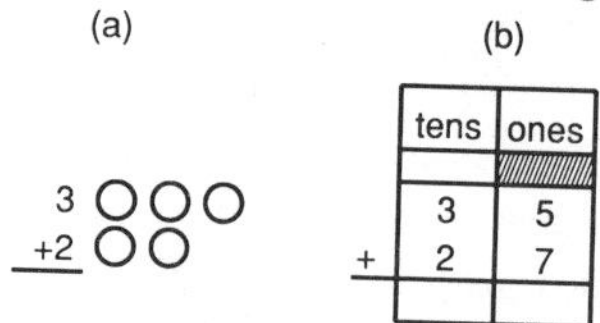

Fig. 3.3. Exercises requiring the writing of correct answers

In a constructivist primary mathematics program that I have been developing in a public school since 1984, we found that second graders who have been encouraged to do their own thinking since kindergarten invent their own procedures for adding multidigit numbers. When given problems such as 35 + 27 written vertically, horizontally, or in any other form on the chalkboard, these second graders invent procedures such as the following:

30 + 20 = 50	30 + 20 = 50
5 + 7 = 12	5 + 5 = "another ten"
50 + 10 = 60	50 + 10 = 60
60 + 2 = 62	60 + 2 = 62

Since children (as well as adults) think of 35 as "thirty and five" rather than as "five and thirty," it is not surprising that they start by adding the tens first. The algorithm now taught at school goes counter to the way children naturally think. If we let them do their own thinking, frames like the one shown in figure 3.3(b) are completely superfluous.

As for the addition of single-digit numbers in first grade, the worksheet approach interferes with the possibility of children's remembering combinations such as "3, 2, 5" and "4, 2, 6" by requiring them to write numerals. By contrast, in a board game using two dice, children are free to think about 3 + 2 in relation to 4 + 2.

The essence of arithmetic should be thinking. First graders who are encouraged to think are likely to change 5 + 6 to (5 + 5) + 1, and 7 + 8 to (8 + 2) + 5, (7 + 3) + 5, or (7 + 7) + 1 if they do not know the answer. Those who are encouraged only to count and write usually count and write, without trying to use the knowledge they already have to figure out the unknown.

Mathematical symbols belong to social knowledge, that is, knowledge based on conventions worked out by people. Children who can think, that is, those who can do the logico-mathematical part of arithmetic, can easily learn to write mathematical symbols. By prematurely focusing on children's learning of symbols and by overlooking the importance of thinking (i.e., constructive abstraction), traditional mathematics instruction teaches children to manipulate symbols on paper instead of constructing logico-mathematical knowledge.

2. *We must encourage children to agree or disagree among themselves rather than reinforce right answers and correct wrong ones.*

A characteristic of logico-mathematical knowledge is that there is absolutely nothing arbitrary in it. Two plus two makes four in all cultures. It follows that in the logico-mathematical realm, children *will* arrive at the truth if they debate long enough. For example, if one child says that 4 + 3 = 7 and another that 4 + 3 = 6, the teacher's role should be to ask other children if they agree rather than reinforce the right answer or correct the wrong one.

Piaget (1950) argued that an exchange of ideas and mutual control (the origin of the need for verification and demonstration) are essential for children's development of logic. Perret-Clermont (1980) and Doise and Mugny (1984) experimentally verified this statement. They found that the children in experimental groups who had a chance to agree, disagree, and convince each other in small groups demonstrated higher-level thinking on the post-test than those in control groups, who did not have this opportunity. Other people are not the sources of logico-mathematical knowledge. However, other people's arguments can cause children to reexamine their own thinking and to construct a higher level of thinking from within.

The preceding discussion has emphasized the importance of encouraging children to express their own opinions honestly. This emphasis on children's autonomous and honest thinking is in contrast with traditional instruction (Piaget 1974). By reinforcing right answers and correcting wrong ones, traditional mathematics instruction unwittingly stifles children's ability to do their own thinking. Children change their minds only when they are convinced that someone else's argument makes better sense.

EXAMPLES OF CLASSROOM ACTIVITIES

The best types of classroom activities entail the use of situations in daily living, group games, and discussions about different ways of solving problems.

The Use of Situations in Daily Living

Situations in daily living are full of mathematical activities when we look for them. In kindergarten, children can be asked to bring just enough cups, napkins, or scissors for everyone at their table. They can also be asked to bring materials from home for art projects, such as empty tubes of toilet paper and empty containers. If, for example, puppets cannot be made until there are enough tubes for everyone in the room, children are motivated to count them each time new ones have been brought from home. In first and second grade, taking attendance and voting become daily occasions for arithmetic, and many second graders become able to count lunch money and help with bookkeeping chores.

The principle of teaching is to ask, "How can we find out?" and to ask children to come to an agreement when there is more than one answer. Children do not need to be motivated artificially in these situations because they usually care about real-life problems.

One day in a first-grade class, for example, the group tried to decide whether or not to continue an activity. When thirteen children voted in favor of continuing it, one child said, "We don't need to vote for the other choice. The thirteen already won." Surprised by the speed and certainty of this child, the teacher asked him how he knew that the thirteen won. He explained: "13 + 13 = 26, and we have only 24 kids today." Most of the other children reacted with blank faces, and the teacher asked them if they were convinced. They were not and said it was necessary to count the people voting for the other choice. After empirically counting the eleven people who voted for the other alternative, everybody was satisfied.

In traditional mathematics instruction, "story" problems are often considered applications of computational skills and are presented after computational exercises. This sequence should be reversed because children construct logico-mathematical knowledge out of daily living. Computation with numbers, which do not involve contexts, should come after a great deal of problem solving with real-life contexts. Many other examples of situations in daily living that a teacher can use are given in Kamii (1982, 1985, 1989).

Group Games

Games are excellent activities because children play them to please themselves rather than the teacher. They are desirable because in games children care about sums, supervise each other, and give immediate feedback. In the worksheet approach, by contrast, feedback is delayed, and young children do not remember or care about what they did yesterday. Games are good also because the social interaction they require contributes greatly to children's social and moral development. In the worksheet approach, children work alone and thus do not develop socially or morally.

The card game war is excellent for kindergarten, and it can be changed to double war in first grade. In war, all the cards are dealt facedown to two players. Each player then turns over the top card of his or her stack, and the person who has the bigger number takes both cards. In double war, each player turns over two cards and compares the total of those two cards to the total of the opponent's two cards.

Dice games, too, can be varied according to children's developmental levels. In kindergarten, it may be best to use only one die with boards consisting of grids. Players roll the die and fill a corresponding number of empty spaces with chips or bottle caps; the first person to fill the board is the winner. The total number of empty spaces may be twenty or thirty in kindergarten, but this number can be increased to fifty in first grade. By first grade, it is good to use two dice for addition and a board that requires players to take as many steps as the total indicated by the two dice. In second grade, this game can be modified to allow moving only if the total is an even number.

Many other examples of similar games can be found in Kamii and DeVries (1980) for kindergarten, in Kamii (1985) for first grade, and in Kamii (1989) for second grade.

Discussions of Problems

Double-column addition involves place value and is particularly difficult for second graders. This is why we introduce a third type of activity in second grade, namely, discussions. The teacher initiates these discussions by putting one problem after another on the chalkboard. For example:

$$\begin{array}{r}8\\+7\\\hline\end{array}\qquad\begin{array}{r}10\\+12\\\hline\end{array}\qquad\begin{array}{r}17\\+\ 6\\\hline\end{array}\qquad\begin{array}{r}23\\+19\\\hline\end{array}$$

The teacher then asks the group for all the different ways they can think of to solve each problem. Refraining from saying that an answer is right or wrong or that one procedure is better than another, the teacher encourages children to explain their solutions and to tell each other whether or not an explanation makes sense to them. Examples of typical procedures children invent have already been given above. Further detail of these discussions can be found in Kamii (1979), Kamii and Joseph (1988), and Kamii (1989).

CONCLUSION

Constructivism implies fundamental changes in the way arithmetic is taught. If we assume on the one hand that children must learn arithmetic by internalization from the environment, we show them how to add, subtract, multiply, and divide, and reinforce right answers in exercises. If, on the other hand, we know that children construct their own logico-mathematical knowledge from the inside, we encourage them to do their own thinking and to exchange viewpoints among themselves. Young children will eventually construct the algorithms that are now prematurely imposed on them. By letting them change their minds only when they are convinced that another idea makes better sense, we encourage them to build a solid foundation that will enable them to go on constructing higher-level thinking. Further implications of constructivism for adaptive mathematics teaching are discussed in the article by Steffe in this yearbook.

REFERENCES

Doise, Willem, and Gabriel Mugny. *The Social Development of the Intellect.* New York: Pergamon Press, 1984.

Inhelder, Bärbel, and Jean Piaget. "De l'Itération des actions à la récurrence élémentaire." In *La Formation des raisonnements récurrentiels,* by Pierre Gréco, Bärbel Inhelder, B. Matalon, and Jean Piaget. Paris: Presses Universitaires de France, 1963.

Kamii, Constance. *Double-Column Addition: A Teacher Uses Piaget's Theory.* Videotape. 1989. Distributed by NCTM and Teachers College Press.

———. *Number in Preschool and Kindergarten.* Washington, D.C.: National Association for the Education of Young Children, 1982.

———. *Young Children Continue to Reinvent Arithmetic, 2nd Grade.* New York: Teachers College Press, 1989.

———. *Young Children Reinvent Arithmetic.* New York: Teachers College Press, 1985.

Kamii, Constance, and Rheta DeVries. *Group Games in Early Education.* Washington, D.C.: National Association for the Education of Young Children, 1980.

Kamii, Constance, and Linda Joseph. "Teaching Place Value and Double-Column Addition." *Arithmetic Teacher* 35 (February 1988): 48–52.

Perret-Clermont, Anne-Nelly. *Social Interaction and Cognitive Development in Children.* London: Academic Press, 1980.

Piaget, Jean. *The Psychology of Intelligence.* 1947. Reprint. London: Routledge & Kegan Paul, 1950.

———. *To Understand Is to Invent.* 1948. Reprint. New York: Viking Press, 1974.

4

The Role of Routine Procedures in the Development of Mathematical Competence

James Hiebert

THE aim of this article is to consider the ways in which carrying out routine procedures contributes to the development of mathematical competence. Routine procedures are those procedures that appear regularly and frequently in school mathematics programs. Prime examples are the paper-and-pencil algorithms for computation. Indeed, learning how to add, subtract, multiply, and divide with speed and accuracy have long been goals of school mathematics. During the past decade, however, a number of well-conceived and highly regarded reports in the mathematics education community have recommended a de-emphasis on the arithmetic paper-and-pencil algorithms and an increased emphasis on more creative aspects of mathematics, such as problem solving (National Council of Teachers of Mathematics 1980; Commission on Standards for School Mathematics 1989). Where does this leave the practice of routine procedures? What role should the mastery of routine procedures play in future mathematics programs?

The article will examine the role of performing routine procedures from a *cognitive* perspective. This means that the discussion will focus on *how* competence develops and on *how* the practice of routine procedures contributes to this development but not on *what* mathematics content students should master. Decisions about what content should be taught are curricular decisions and must take into account noncognitive as well as cognitive fac-

Acknowledgments are gratefully extended to Thomas J. Cooney, William B. Moody, Diana Wearne, and the Editorial Panel for reviewing an earlier draft of this article and to the National Science Foundation (Grant No. MDR 8651552) for their partial support while it was written. The opinions expressed, however, are those of the author rather than the reviewers or the Foundation.

tors. The interest here is examining the cognitive processes involved in executing routine procedures and the cognitive effects of such experience on students' mathematical competence.

LEARNING MATHEMATICS MEANINGFULLY

To set the stage for considering the cognitive processes involved in performing routine procedures, we will find it useful to think about what it means to learn mathematics with understanding or with meaning. A theme that runs throughout the article is that understanding is essential in the development of competence. There are a number of classic writings within the mathematics education literature about learning with understanding (e.g., Brownell 1935, 1947; Bruner 1960; Fehr 1955; Van Engen 1953). Although it would be impossible to capture here all that has been written in these articles and others about meaning and understanding, it is possible to describe one view of learning with understanding that both is consistent with most writings and provides a context for considering the contribution of practicing routine procedures to such learning.

To think or talk about mathematical ideas, we must represent these ideas in some way. This may be obvious, but it is an important point. Suppose you are teaching first graders about simple addition and subtraction. To communicate the ideas to the students, you need to choose a way of representing the ideas. You will probably talk with the students and try to represent the ideas in words; you might use concrete objects or draw pictures to demonstrate the ideas; you might even set up a real-life scene and place the student in a familiar adding or subtracting situation; eventually you will probably use written symbols to represent the ideas (e.g., $5 + 2 = 7$). These are five common ways in which we represent mathematical ideas: spoken language, concrete objects, pictures, real-life situations, and written symbols (Lesh, Post, and Behr 1987). All the representations are useful. The context and purpose determines which representation is most appropriate. But note: Meaning or understanding in mathematics comes from building or recognizing relationships either *between* representations or *within* representations.

Building Relationships between Representations

Suppose that Martha is in first grade and is learning about addition and subtraction. She is working on a story that says she received three birthday gifts from her friends and four gifts from her family. To figure out how many gifts she received in all, she uses blocks to stand for the gifts, forms a set of three and a set of four, and then counts them all. She then learns that the situation can be recorded on paper as $3 + 4 = 7$. As Martha engages in numerous similar activities, she is constructing relationships among at least three forms of representation—real-life situations, concrete objects, and written symbols. It is in constructing relationships among the representations that Martha develops meaning for addition. Constructing these relationships is not an easy thing to do and often takes some time. Although teachers can help students see connections or relationships between and among representations by pointing out how they capture the same ideas (see Hiebert and Lindquist [1990] for some suggestions), students must build the relationships for themselves.

Relationships between different representations generate meaning for children from their earliest encounters with mathematics through all the years of formal schooling. Even more advanced mathematical ideas can be represented in the five representation forms identified earlier. For example, algebraic equations that are usually expressed with written symbols can be represented through a real-life situation or depicted in a graph. Establishing connections between these representations is as important at this later point in a student's career as it is during the first years of school.

Building Relationships within Representations

Meaning in mathematics can also be developed by relating items within a particular form of representation. Often this process involves noticing patterns or regularities within the representation system. For example, suppose base-ten blocks are used to represent decimal fractions. The large cube can be assigned the unit value, or 1. Values of the remaining blocks are then determined. The flat block represents one-tenth, the stick becomes one-hundredth, and the small cube becomes one-thousandth. When students first encounter decimal fractions, they may believe that the decreasing fraction values end with thousandths, the smallest block. But if they recognize the pattern of *repeated* partitioning by 10 and the corresponding decrease in the size of the blocks, they may realize that the blocks could represent smaller and smaller fraction values, if one had a thin enough saw to cut the blocks. Recognizing this pattern provides new meaning for the blocks as a representation of decimals. The recognition of the pattern also gives new information about decimals themselves—information that can be connected from the blocks to other forms of representation.

Patterns and regularities occur in all representation forms, and meaning is generated as the patterns are recognized within any of the five forms. As patterns and regularities are recognized, the representation form or system takes on an important consistency and structure. Consequences of actions in the system become predictable and make sense. Written symbols are a particularly important form of representation in school mathematics. Considering how meaning is developed within systems of written symbols brings us to the main question: How do the cognitive processes involved in executing routine procedures contribute to learning mathematics meaningfully?

COGNITIVE PROCESSES INVOLVED IN MASTERING ROUTINE SKILLS

There are two very different kinds of cognitive processes that can be involved in executing routine procedures. One is a process called *automatization.* Procedures are practiced over and over until they can be executed automatically without thinking. The other process can be called *reflection.* As a procedure is executed, the person consciously thinks about what is happening and why it is happening just that way.

Automatizing Routine Procedures

Teachers are probably familiar with activities that promote the automatization of procedures. Suppose the procedure is the multiplication algorithm for two-digit numbers, such as 37×42. To automatize the procedure, students need to perform many exercises like this over an extended period of time. The more exercises that are practiced, the better students get at multiplying quickly and accurately. Even if students do not completely automatize the procedure, repeated practice results in more efficient execution.

Efficient execution of a procedure is the key to understanding the cognitive benefits of automatization. The more efficiently a procedure is executed, the less mental effort required. If a procedure is fully automated, it can be run off without thinking, without any mental effort.

Consider the simple, one-step procedure of multiplying two single-digit numbers. If the multiplication facts have been practiced enough, the product of two numbers (e.g., 8×7) can be given automatically. Little mental effort is needed to produce the answer. If the multiplication facts are not memorized, the product can still be found. But the strategies that are usually used, many of them involving counting in some way, all require considerable effort.

The significance of reducing mental effort is that all of us have a limited amount of mental effort that can be expended at a given time (Case 1985). If we use all the effort to complete one task, we may have nothing left with which to think about another task. However, if we have automatized a procedure for completing a task, then executing the procedure takes little effort, and we can solve another task at the same time. Our limited mental effort is free to think about the second task while the procedure for the first task is running on its own.

Reducing mental effort may be advantageous, but it does not contribute directly to the development of meaningful mathematical knowledge. What it does is free up effort that can be used in another way. Perhaps the saved effort will be used to search for relationships, to create meaning. If it is, then the automatization of procedures can contribute *indirectly* to the development of mathematical competence, but there is no guarantee that students will use the saved effort in this way.

The automatization of procedures can be justified only when they are required in the service of completing other tasks. To illustrate the benefits of the automatization of procedures, consider the problem of determining the weight of a class aquarium when it is filled with water (should it be filled in the rest room and carried to the classroom or will it be too heavy?). Suppose Richard, a student in the class, is given the task of estimating the weight. One strategy is to find the capacity of the aquarium (it is found to measure 49 cm × 31 cm × 20 cm), find how many smaller units of capacity it will hold (e.g., liters), and then, using the weight of the smaller unit, estimate the weight of the aquarium. This strategy, as well as any other, requires a good deal of computational estimation. If the multiplication number facts are automatized, along with some rules for multiplying powers of 10, the problem becomes one of thinking about, and following, the most efficient strategy. Carrying out the strategy is relatively simple and quick. If the multiplication facts are not automatized, the problem may become a laborious series of paper-and-pencil computations. It is likely that Richard would get bogged down in working through the details of the algorithms. In fact, an estimate would not be much easier for Richard than a precise result.

The availability of calculators and computers certainly will affect the answer to the question of which skills to automatize. The benefit of automatizing procedures is to reduce the mental effort required to execute them, and electronic calculators reduce the mental effort to almost zero. If calculators are readily available, very few calculation skills would need to be automatized. A plausible suggestion is that for the sake of convenience, the whole-number arithmetic facts along with the base-ten notation rules for combining larger and smaller numbers (e.g., $200 \times 40 = 8000$; $0.2 \times 0.4 = 0.08$) should be automatized. These facts and procedures provide benchmarks that permit access to numerous sums and products and enable useful estimation skills to develop.

One final comment must be made about the repeated practice of routine procedures. It is a popular belief that frequent and repeated practice is necessary for students to remember the procedures. There is no question that for simple, one-step procedures (e.g., addition and multiplication facts), practice promotes quick recall (Thorndike 1922). But for more complex algorithms, such as computations with multidigit numerals, fractions, or algebraic expressions, it is not clear that large amounts of practice are necessary or even the best way to promote recall. Procedures like these probably are remembered better if time is spent making sense of them (Hiebert and Lefevre 1986; Skemp 1978). If we want students to remember procedures, we should ask them to step back and think about the procedures they are using rather than practicing more exercises. This brings us to the second kind of cognitive process that can be involved in executing procedures.

Reflecting on Routine Procedures

It was suggested earlier that one way of developing meaning for mathematics is recognizing patterns and regularities *within* a particular form of representation. The process of reflection is exactly the kind of cognitive process that is able to search for patterns and regularities in the system. Reflecting on procedures means considering in a deliberate, conscious way the patterns that may emerge as procedures are carried out. The procedures of interest are procedures with written symbols, so the process of reflection is a process of searching for relationships—patterns and regularities—within the system of written symbols.

Because reflection is a very different process from automatization (the process that is active during popular drill-and-practice worksheet activities), it may be useful to consider several examples. Consider again the problem 37×42 and the multiplication algorithm for performing the task. We usually ask students to spend a considerable amount of time practicing the paper-and-pencil algorithm on exercises like this one. What would we do if we wanted them to reflect on the procedure instead of just working toward automatization? The conventional form of the procedure (shown in fig. 4.1a) is quite compact and hides many of the patterns that exist. Although students might discover some patterns if they worked through the algorithm slowly and thoughtfully, teachers can help them by posing appropriate questions and suggesting alternative formats for recording results. The elaborated format in figure 4.1b records the information so that patterns are more accessible. For instance, relationships to single-digit multiplication are more

(a)	(b)	(c)	
37	37	37	
42	42	42	
74	14	14	2 × 7
148	60	60	2 × 30
1554	280	280	40 × 7
	1200	1200	40 × 30
	1554	1554	

Fig. 4.1. Alternative formats for recording the multiplication algorithm

evident, and new relationships, such as the patterns of multiplying numbers increased by powers of 10, can be discovered. The further elaborated format in figure 4.1c makes these patterns and relationships even more obvious. These elaborated formats are not suggested just as a prelude to a tightening of procedure (a) and its repeated practice. The elaborated formats are appropriate, *in themselves,* as aids toward achieving a different goal. The goal here is not to reach efficient execution but to see relationships—patterns and regularities—that help the algorithm, and the entire system, make sense.

A second example of recognizing patterns involves the traditional division algorithm and junior high school students who are working a number of division exercises with calculators. They are changing common fractions to decimal form. The teacher asks if they notice any patterns in the results. Malcolm suggests that the decimal expression always ends in 0. Henrietta doesn't agree and shows that the expression for 5/3 has a 6 that keeps repeating. Malcolm thinks that the expression will eventually end in 0. Susan says that she has a fraction whose decimal form does not repeat or terminate: 3/17. After some discussion, the teacher suggests that they look at the algorithm that is producing these results and search for clues that might help them see what kinds of decimal expressions to expect. As the algorithm is carried out slowly on paper for several different examples, students begin noticing that the number of different differences for each subtraction step is limited by the size of the divisor. They then notice the patterns that are produced in the workspace when a difference is first repeated. These repeated cycles of calculation in the workspace are then connected to the terminating or repeating nature of quotients. An important point of this example is that the division algorithm is not used here as an efficient computational device but rather as a procedure that can be analyzed deliberately to reveal new insights into the structure of the system.

A final example comes from more advanced mathematics in high school. A frequent topic in second-year algebra is the behavior and use of exponents. Students become familiar with the fact that when like bases are multiplied, the exponents are added (e.g., $2^3 \times 2^5 = 2^8$). Of course, students can detect this pattern rather than being told the rule if they are engaged in appropriate activities and asked to reflect on the regularities they see. But assume that they recognize, for whatever reason, the behavior of exponents when multiplying like bases. Now they are faced with a problem such as $64^{1/2} \times 64^{1/3}$. How might one express this number in a simpler way? Students who have confidence in the regularities of the symbol system might predict $64^{1/2 + 1/3}$, or $64^{5/6}$. The prediction can be checked in a variety of ways, but the process of recognizing patterns in simple cases and using the patterns to make predictions about new problems is an important mathematical process, a process that can contribute to the development of meaning and understanding. Reflecting on routine procedures can support this process.

The written symbol systems of mathematics are full of patterns. Recognizing these patterns is a way of building relationships, relationships that reveal the structure and consistency and elegance of the systems. Performing routine procedures provides an entrance into this world of patterns, but only if the goal is to reflect on the procedures rather than to practice them.

Treating routine procedures reflectively introduces another consideration into curricular decisions. Whether to study a routine algorithm should be decided not only on the benefits of automatizing the algorithm but also on the algorithm's potential to reveal patterns in the system, for recognizing patterns does much more than help students make sense of the algorithm; it helps students make sense of the system.

A SUGGESTED SEQUENCE FOR DEVELOPING MEANING AND PRACTICING PROCEDURES

Three different kinds of cognitive processes have been described: constructing relationships between forms of representation, constructing relationships within forms of representation (searching for patterns), and automatizing procedures. The last two are directly involved in performing routine procedures in mathematics. A practical question for the teacher is where to start. How should activities be sequenced to maximize the benefits that can be derived from engaging in these processes? A conclusive answer to the question cannot yet be given, but we do know enough to offer some informed suggestions.

When students encounter a new symbol system, it appears that the most beneficial sequence of activities is one that engages the cognitive process of building relationships between symbols and *other* representation forms before searching for patterns within the symbol system or practicing procedures (Hiebert 1988; Mason 1987). Suppose fourth graders are encountering decimal fractions for the first time. Initial activities should assist students in establishing connections between decimal fraction quantities represented in a familiar or salient form, such as base-ten blocks, and the standard written symbols. Such connections inject appropriate meaning into the written symbols (Van Engen 1949). Once students have developed meaning for the symbols, they are in a good position to develop simple procedures, sometimes on their own. Adding and subtracting decimals, for example, can be understood as combining quantities that are alike (instead of following a meaningless "line up the decimal points" rule), and decimals can be ordered simply by thinking about the quantities they represent. As procedures are developed, students can be asked to reflect on patterns within the symbol system that emerge from applying the procedures on a variety of problems.

Drill-and-practice activities should be prescribed *after* meaning has been developed by the students and after some careful thought by the teacher. Some evidence suggests that if skills are practiced too soon, it is more difficult for students to go back and develop the meaning that should be there initially (Brownell and Chazal 1935; Resnick and Omanson 1987; Wearne and Hiebert 1988). Remember also that automatizing procedures in itself does not contribute to developing meaningful mathematical knowledge. Automatized procedures are helpful only if the mental effort that would otherwise be used to execute them is directed toward more meaningful tasks.

CONCLUSIONS

The mastery of routine procedures consumes a major share of mathematics lessons in today's schools. With the recent recommendations to decrease the emphasis on common drill-and-practice activities, it is worthwhile to examine the contributions of performing routine procedures to the development of meaningful and useful mathematical knowledge. One way of examining these contributions is to consider the cognitive processes involved in carrying out the procedures. Two processes stand out. One is the process of automatization. Automatizing procedures involves repeated practice with a goal of achieving efficient execution. Because this process does not contribute directly to the development of meaningful knowledge, drill-and-practice activities are appropriate only when the automaticity of a particular skill is essential for completing more meaningful tasks. The second process is one of reflection. Reflecting on how and why procedures work as they do leads to a recognition of patterns and regularities within the system. Because the search for patterns is one way to create meaningful knowledge, reflection on procedures is a process that should be encouraged.

REFERENCES

Brownell, William A. "The Place of Meaning in the Teaching of Arithmetic." *Elementary School Journal* 47 (1947): 256–65.

________. "Psychological Considerations in the Learning and Teaching of Arithmetic." In *The Teaching of Arithmetic*. Tenth Yearbook of the National Council of Teachers of Mathematics. New York: Teachers College, Columbia University, 1935.

Brownell, William A., and C. B. Chazal. "The Effects of Premature Drill in Third-Grade Arithmetic." *Journal of Educational Research* 29 (1935): 17–28.

Bruner, Jerome S. "On Learning Mathematics." *Mathematics Teacher* 53 (December 1960): 610–19.

Case, Robbie. *Intellectual Development: Birth to Adulthood*. New York: Academic Press, 1985.

Commission on Standards for School Mathematics. *Curriculum and Evaluation Standards for School Mathematics*. Reston, Va.: National Council of Teachers of Mathematics, 1989.

Fehr, Howard F. "A Philosophy of Arithmetic Instruction." *Arithmetic Teacher* 2 (April 1955): 27–32.

Hiebert, James. "A Theory of Developing Competence with Written Mathematical Symbols." *Educational Studies in Mathematics* 19 (1988): 333–55.

Hiebert, James, and Patricia Lefevre. "Conceptual and Procedural Knowledge in Mathematics: An Introductory Analysis." In *Conceptual and Procedural Knowledge: The Case of Mathematics*, edited by James Hiebert, pp. 1–27. Hillsdale, N.J.: Lawrence Erlbaum Associates, 1986.

Hiebert, James, and Mary M. Lindquist. "Developing Mathematical Knowledge in the Young Child." In *Teaching and Learning Mathematics for the Young Child*, edited by Joseph N. Payne. Reston, Va.: National Council of Teachers of Mathematics, 1990.

Lesh, Richard, Thomas Post, and Merlyn Behr. "Representations and Translations among Representations in Mathematics Learning and Problem Solving." In *Problems of Representation in the Teaching and Learning of Mathematics*, edited by Claude Janvier, pp. 33–40. Hillsdale, N.J.: Lawrence Erlbaum Associates, 1987.

Mason, J. H. "What Do Symbols Represent?" In *Problems of Representation in the Teaching and Learning of Mathematics*, edited by Claude Janvier, pp. 73–81. Hillsdale, N.J.: Lawrence Erlbaum Associates, 1987.

National Council of Teachers of Mathematics. *An Agenda for Action*. Reston, Va.: The Council, 1980.

Resnick, Lauren B., and Susan F. Omanson. "Learning to Understand Arithmetic." In *Advances in Instructional Psychology*, vol. 3, edited by Robert Glaser, pp. 41–95. Hillsdale, N.J.: Lawrence Erlbaum Associates, 1987.

Skemp, Richard R. "Relational Understanding and Instrumental Understanding." *Arithmetic Teacher* 26 (November 1978): 9–15.

Thorndike, Edward L. *The Psychology of Arithmetic*. New York: Macmillan, 1922.

Van Engen, Henry. "An Analysis of Meaning in Arithmetic." *Elementary School Journal* 49 (1949): 321–29, 395–400.

———. "The Formation of Concepts." In *The Learning of Mathematics: Its Theory and Practice*, Twenty-first Yearbook of the National Council of Teachers of Mathematics, edited by Howard F. Fehr, pp. 69–98. Washington, D.C.: The Council, 1953.

Wearne, Diana, and James Hiebert. "A Cognitive Approach to Meaningful Mathematics Instruction: Testing a Local Theory Using Decimal Numbers." *Journal for Research in Mathematics Education* 19 (November 1988): 371–84.

5

Adaptive Mathematics Teaching

Leslie P. Steffe

EVERY mathematics teacher has probably been asked a question similar to the one an anguished, freckle-faced fourteen-year-old student in first-year algebra once asked me: "Why do I have to learn this junk? I'll never use it!" I have never forgotten the question because it was such a shock at the time. Another comment made by a sixteen-year-old in second-year algebra was every bit as penetrating: "You just don't understand!" At the time, I had recently spent five years studying mathematics and science, so I fancied myself a well-prepared mathematics teacher. Yet here were my students asking me to understand them and to teach mathematics in such a way that it would be useful to them! Those comments haunted me for a long time because those two students were representative of most of the 150 or so mathematics students I taught each year. It became increasingly clear to me that "algebraic reasoning" did not serve them in the organization of their experiential world, nor did they see how it might be used in that way. Moreover, they did not accept algebraic reasoning as belonging to them, as being an important part of their concept of self. Whatever their answer to the question "Who am I?" might have been, it did not include the algebra I was teaching. In general, algebra belonged to me as their mathematics teacher and to other "people of mathematics" (Steen 1986).

Of course, it is always possible that what I learned is particular to my personal experiences as a mathematics teacher. However, on the basis of reports concerning education over the past decade, what I learned seems to be shared by other mathematics teachers as well (e.g., National Council of Teachers of Mathematics 1989; National Research Council 1989; National Science Board Commission on Precollege Education in Mathematics, Science, and Technology 1983). A comment made by the authors of *Everybody Counts* especially reinforces my beliefs (National Research Council 1989): "To improve mathematics education for all students, we need to expand

I would like to thank the reviewers of this yearbook for their comments on earlier drafts of this article. Special gratitude is expressed to Thomas Cooney for his encouragement.

teaching practices that engage and motivate students as they struggle with their own learning. In addition to beckoning with the light of future understanding at the end of the tunnel, we need even more to increase illumination in the interior of the tunnel" (p. 58).

EXPANDING TEACHING PRACTICES

How teaching practices might be expanded so that our students experience mathematics as being useful to them and accept their developing power of mathematical reasoning as being an essential part of their concept of self is not immediately obvious. Would it help us mathematics teachers to emulate the teaching practices of our mathematics teachers? Most of us have certainly found mathematics to be intrinsically interesting, and our concept of ourselves as being mathematical has been profoundly influenced by our mathematics teachers. Before teaching precollege mathematics, I had studied with seven mathematics professors (thirteen more were to come). To my knowledge none of them had ever considered how mathematics might be useful to me as a student beyond preparing me for the next course or level of study, nor had they ever tried to understand how I might think and thus take that into consideration in their teaching. As a student, I always understood that it was I who had to adapt and learn to think like my professors. I never seriously considered, therefore, that understanding the mathematical knowledge of my students should be a part of my responsibility.

A Basic Responsibility

Some very able mathematicians were among those initial seven professors, and I had built up my way of teaching by emulating what I thought were excellent role models. My basic responsibility as a mathematics teacher, as I had inferred it from observing my professors, was to explain mathematics to my students so they would understand the mathematical concepts and relations rather than just learn the mechanics of doing mathematics. And the current reports on education notwithstanding, I believe that is a responsibility most teachers accept. Mathematics teachers generally resort to emphasizing the mechanics of doing mathematics out of desperation when attempting to teach an extant syllabus to 150 or so mathematics students a day.

As a precollege mathematics teacher, I intended to play the role of a mediator between my students and mathematics and to reveal mathematics to them as a meaningful set of relationships that would have relevance to them as they organized their experiences and their self-understanding deepened. The students, however, were to learn mathematics as it is, not as they might make it to be. So, at the time, understanding the mathematics of my students and how it might be useful to them was not part of what I took to be my responsibility as a mathematics teacher. I now take a different view of that responsibility, and to clarify the difference, I use one of my favorite examples.

Explaining the Sum of the First *n* Natural Numbers

When teaching arithmetic series, I started by explaining to my students how to find the sum of the first *n* natural numbers. After explaining how to find the sum of the first few natural numbers in a way that would anticipate the general method I wished to demonstrate, I presented the following general method:

$$\begin{array}{rcl} S & = & 1 + 2 + 3 + \cdots + (n-2) + (n-1) + n \\ S & = & n + (n-1) + (n-2) + \cdots + 3 + 2 + 1 \\ \hline 2S & = & (n+1) + (n+1) + (n+1) + \cdots + (n+1) + (n+1) + (n+1) \end{array}$$

After writing the sum of the first *n* natural numbers and calling that sum *S*, I explained that they could be written in reverse order and then the two series could be added term for term. So, because there are *n* terms named $(n + 1)$, twice the sum is the number of terms times the number of terms increased by 1. That is, $2S = n \times (n + 1)$ or $S = n \times (n + 1)/2$. To me, this was an elegant way of teaching that stressed the mathematical relationships that the formula symbolized. My goal was for my students to understand the formula; I seldom asked them to simply remember it, but instead to understand the operations it symbolized.

Following a Demonstration

Following a demonstratiaon means being able to produce the steps that compose it. In this sense, a majority of the students could follow the demonstration and could understand its logic *after the fact.* This, along with using the formula, certainly could be called mathematical experience, but it was not those experiences that encouraged intuition, insight, or creativity. Following a demonstration or using a formula did not lead to the knowledge and confidence on the part of the students that they could engage in the activity that yields mathematics. They did not view themselves as being participants in producing mathematical results in the mathematics classroom and soon learned to rely on me, and sometimes on their textbook, as the source of the mathematics that they were supposed to learn. They knew the teacher would tell them the basic principles, so why should they bother with the hard work that is involved in doing mathematics or with even reading the textbook?

A Change in Basic Responsibility

As mathematics teachers, we know that the fundamental characteristic of the individual who chooses to stay in mathematics is the ability to engage in productive mathematical reasoning rather than simply follow demonstrations and use formulas. But, following my mathematics professors, I believed that such reasoning is confined only to the mathematically talented and expected it only of a few of my "best" students. I now realize how discriminatory this is and how it excludes almost all students from mathematics. Even more important is that I now realize that mathematical operations can be a constitutive part of the development of the human mind and that what it means to understand our students is to understand those operations of the mind, how they might evolve, and their products.

So, I now consider that understanding the mathematical operations of my students and how my students might construct those operations in the context of doing mathematics in the classroom is a part of my responsibility as a mathematics teacher. For me, this is the essence of what distinguishes a mathematics teacher as a professional. Mathematics teachers must take responsibility for understanding the mathematics of their students and how their students learn mathematics in the context of the ongoing mathematics classroom (Steffe 1988). Nothing less is acceptable.

When mathematics teachers perceive that they are saddled with an a priori mathematics curriculum that they are obligated to teach, mathematics teaching either proceeds in a way that is similar to what I have described above or else focuses on the mechanics of doing mathematics. In both situations, it is essentially nonadaptive. As mathematics teachers, we tend to assume that it is the students who must do the adapting and must learn to think as we do. We essentially do not teach to learn our students' ways and means of operating; rather, we teach so our students learn our ways and means of operating. But neither is sufficient, because we each must learn the thinking of the other. Emphasizing adaptation on the part of teachers and students as suggested above involves a change in paradigm from stressing mathematics as being the way it is, independent of human beings, to stressing the mathematical knowledge of the students and the teachers.

THE MATHEMATICAL REALITIES OF CONCERN TO TEACHERS

Because our experiential world (including our conceptual world) is all we can know, the question of whether our mathematical knowledge can ever match that of another cannot be decided. Because of the subjectivity of experience, all we can do is interpret the language and actions of the other person with whom we are working using our own conceptual operations in an attempt to understand what the reality of the other person might be like. As perceived by a teacher, then, a student's mathematical reality is an invention of the teacher, and the relation is reciprocal. So, through interpreting each other's language and actions, a teacher and a student can formulate a negotiated reality, which is based on the principle of decidability (Born 1965, p. 172). Through performing tests in interactive communication, they can decide if what they mean or what they see is compatible with what each other might mean or see.

As mathematics teachers, then, we have the responsibility of learning the mathematical knowledge of our students through interpreting their mathematical language and actions and on that basis making decisions about what their current mathematical concepts and operations might consist of and about what they might learn. These decisions are a crucial part of teaching and are a means whereby we can harmonize our teaching with our students' learning.

ADAPTIVE MATHEMATICS TEACHING: AN ILLUSTRATION

A change in paradigm from school mathematics as a mind-independent reality recorded in textbooks to mathematics as a human activity that is carried out in the social context of the classroom shifts the focus of mathematics teaching from a process of transferring information to students to interactive mathematical communication in a consensual domain of mathematical experience. When interacting mathematically with students, teachers will find it helpful to remember that language is not a means of transporting conceptual structures from teachers to their student, but rather a means of interacting (von Glasersfeld, forthcoming).

I return now to the problem of teaching students to find the sum of the first n natural numbers. But my problem now is very different from before. Now my problem is to create situations for my students in which the mental operations involved in transforming the sum of the first n natural numbers into a product might evolve as a result of "mathematizing," which is the process of organizing experience in ways that are distinctly mathematical. According to Treffers (1987), mathematizing should occur in experiential situations so there will be a chance for students to see how mathematics can be relevant in their lived-in world. The song "The Twelve Days of Christmas" is an example of a culturally embedded situation that students can use their understanding of the natural number sequence to comprehend. If it occurs to them (or if they are asked) to find the number of gifts mentioned for each of the twelve days, this could create a starting point for the construction of the sequence of sums of the first twelve natural numbers. The students are most likely to operate recursively to generate the partial sums, so a general goal should be to generate a shorter method to find how many presents are mentioned on any given day. In the two situations that follow, the students should understand that all the activities in which they are engaging help achieve this general goal.

Facilitating Situations

How students organize a situation can provide a window through which their mathematical concepts and operations can be "seen." However, what is "seen" may not exist for some students apart from the particular occasion of observation. To help my students "lift" the results of their operations from transient and specific experiences and become aware of how those operations are composed, I encourage them to reenact (not practice) their experience—that is, to explain to themselves as well as to others how they organized the situation. However, I have found reenacting past experience and focusing attention on its structure to be impossible for some students and unproductive for others. So, rather than expect the achievement of certain goals in a given situation to involve the construction of a more or less permanent mathematical structure, analogous situations are created and presented to the students.

Some students persist in organizing each new situation independently of how they organized prior situations. That is, they don't make the abstractions that are necessary to construct a permanent mathematical structure. In these situations, I seek problems for the students to organize that are structurally similar to the ones already worked on but where the students can reflect on the conceptual material they use in organization. If I am successful, I encourage the students to then interpret what they have done in the previous situations using their newly constructed structure. I call such structures "instruments of abstraction."

Two situations are discussed below to illustrate these ideas. As students work in such situations, I alert them to be on the lookout for a solution of their "general" goal. But they should learn to accept that they may not see how achieving their immediate goals helps them to achieve their general goal.

The famous handshake problem. Asking a group of, say, five students to find how many handshakes occur when they shake hands with each other exactly once may not seem related to the twelve days of Christmas, but that is why I chose it. Allowing the students to organize the process of shaking hands can be fun as well as a possible source for achieving the general goal. Initially, it usually produces trial and error and an unsystematic way of proceeding. If, however, the students do organize the handshakes so that each of the five people shakes hands four times, they might interpret the situation as multiplicative and express the total number of handshakes as the product of 5 and 4 (Fischbein et al. 1985). The teacher can play a critical role by asking appropriate and timely questions without imposing a structure on the situation that hasn't originated from the students. One question may be whether any pair of students shook hands twice. It may take some discussion among the students for anyone to see that each handshake has been duplicated, so there are $(5 \times 4)/2$ unique handshakes. Failure to see this duplication is usually a major stumbling block that cannot be eliminated by some students.

Another role of the teacher is to reopen situations that have already been organized by students and encourage them to organize these situations in a different way. For those students successful in the situation above, the teacher might ask, "Can we find a way of shaking hands so that no one shakes hands twice?" If each student sits down after shaking hands, then there would be $4 + 3 + 2 + 1$ total handshakes! Thus, the successful students might see that $(5 \times 4)/2 = 4 + 3 + 2 + 1$ without calculating.

Even if they "see" the equality because the two expressions represent the same number of handshakes, the teacher should encourage the students to reenact the handshakes and try to convince themselves that the handshakes eliminated by each student sitting down are identical to the handshakes thrown out when dividing by 2. This encourages students to reflect on the results of their operations and to become aware of the structure of operating, which is crucial for their mathematical progress.

The staircase problem. Although our most talented students might become aware of the structure of operating and see how to achieve their general goal, organizing the handshake problem may have little in common with the twelve days of Christmas for other students. Also, some of the students may not understand the organization of the handshake problem for any one of several reasons. As a teacher, it is critical to have another problem that is structurally isomorphic to the handshake problem. For example, a problem like the following might be posed (Wertheimer 1959, p. 108):

> A staircase is being built along the wall of a new house. It has nineteen steps. The side away from the wall is to be faced with square carved panels of the size of the ends of the steps. The carpenter tells his apprentice to fetch them from the shop. The apprentice asks, "How many shall I bring?" "Find out for yourself," rejoins the carpenter.

To clarify the problem, the students should be encouraged to draw the staircase as in figure 5.1. By using their calculator or simply by counting, the students might find the sum of the first nineteen natural numbers and thereby solve the problem. In fact, that process should be encouraged by the teacher because it gives the students the confidence that they have a way to solve the problem. If, after finding the sum, the students do not want to search for another method, they might be asked to solve the problem of, say, a staircase with 100 steps. Although they still might want to count or use their calculator, either way would be tedious if they could not write a program for their calculator to find the sum. This fact could reveal to the students why it is important to have a short-cut method if it is possible to find one. One reason a student should have for solving a problem is to find a way to reach a goal that is more useful than a currently known method. Problem solutions should constitute functional algorithms for the students that are modifications of other, less efficient algorithms.

Reorganizing a perceptual field is one thing that encourages insight (Wertheimer 1959). One way to reorganize the diagram in figure 5.1 is to move the three squares in the top two rows so that they fill in the missing spaces on the right side of the figure. See figure 5.2. If this is done, the successful students might reorganize the sum of the first five natural numbers as a multiplicative structure of three rows with five items per row, or the product of 3 and 5. But reorganizing a perceptual field in a particular way might not occur to very many students because they have no current notion of where they are going. One possible teaching technique is to ask students to solve problems that specify the operation the students should use to solve the problem but leave for the students' consideration how the solution process should be organized.

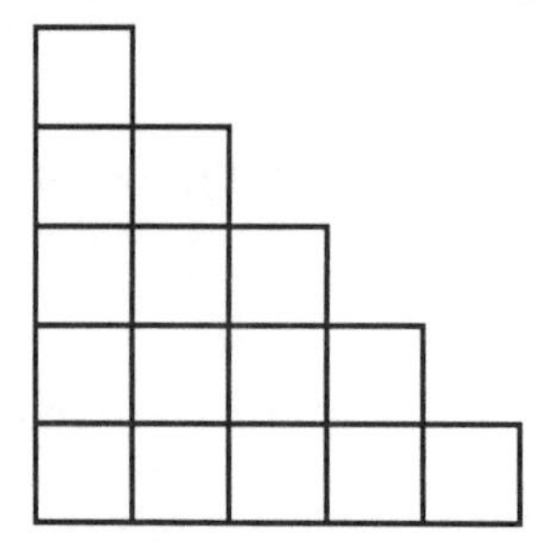

Fig. 5.1. The end of a staircase with five steps

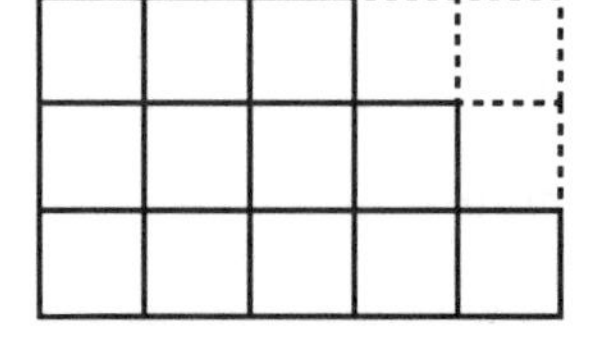

Fig. 5.2. Staircase reorganized

A double staircase problem. As teachers, we have to try to imagine a problem where the solution won't require a major reorganization of students' current ways and means of operating, but one that still engages them in productive mathematical thinking. For example, the teacher might pose a problem where students can fit pieces of a puzzle together—a double staircase problem like this:

> A double staircase is being built for students to walk on at graduation. It has seven steps on a side. The side facing the audience is to be faced with square panels of the size of the ends of the steps. The carpenter building the staircase wants to find how many panels he would need by using the fastest method possible. His idea is to fit the two staircases together on top of one another and then use multiplication. Can you

find the carpenter's method of solving the problem and determine how many panels he needs? (See fig. 5.3.)

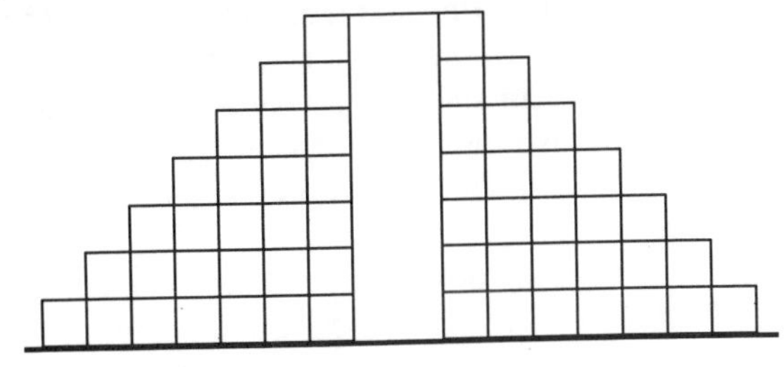

Fig. 5.3. A double staircase with seven steps

In Search of a Common Structure

Intentionally investigating structural similarity is one of the most difficult parts of mathematics learning because it involves becoming aware of the structure of operating—an awareness that has been left to only the most talented students. Resorting to telling or demonstrating the structure of actions might help students to "see" the structure, but that is self-defeating in practice because in the long run it destroys their confidence in their mathematical abilities. Making a table of values for the handshake or the staircase problem (one or both may be solved), where the number of people or the number of steps are varied across problems for the first few natural numbers, may help some students to isolate a solution method as well to see how what they have done helps them to solve the unsolved problems, including that of the twelve days of Christmas.

If making a table is not sufficient to encourage abstraction, finding how many ways two people can be elected cochair from a slate of seven nominees might lead to the abstraction of a permanent mathematical structure because a majority of adolescents can systematically find the permutations of a sequence of natural numbers (Inhelder and Piaget 1958). The same organizing ability is involved in solving the current problem. My goal is for my students to organize their solutions in two ways that they have already met but with different meanings. The first is to list the numbers 1–7 (assuming the candidates are numbered) and then to proceed as shown in figure 5.4. If the first person listed is elected, then any one of the second through seventh could also be elected, so we have six possibilities in this event and a total of $6 + 5 + 4 + 3 + 2 + 1$ different possibilities.

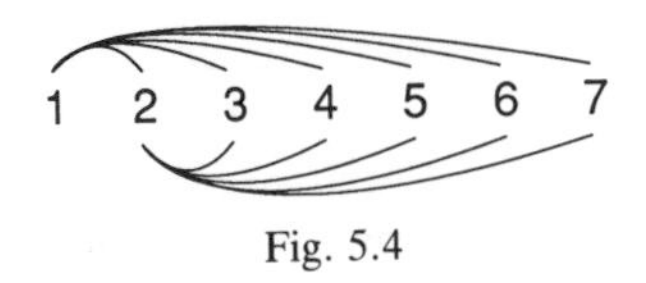

Fig. 5.4

A second way to solve the problem is to list the numbers 1–7 twice (fig. 5.5). Because these two lists of numbers represent the same people, we cannot pair, say, 1 with itself. So we systematically pair 1 in the top list with all the other numbers in the bottom list except itself, and so on. In total, there are 7 × 6 ways of accomplishing the pairings; eliminating duplicates leaves (7 × 6)/2 unique pairs.

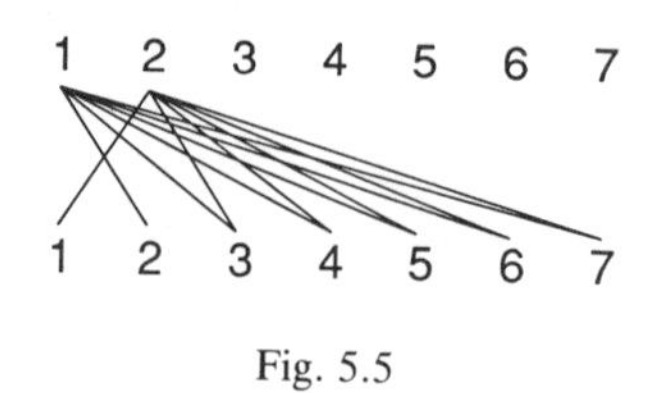

Fig. 5.5

These pairings use the number sequence as conceptual material, not experiential material like handshakes or rows of panels. Because of the freedom from such experiential material and because the students can reflect on their solution method in this exercise, they can profitably be asked to reflect on the two ways of proceeding and isolate the way in which they produce exactly the same pairs, and the multiplicative and additive structures of operating. They should now be asked to relate what they have just done to the handshake and staircase problems in an attempt to isolate the corresponding methods.

PERSPECTIVES

A change in the world view of mathematics teachers is a necessary part of major reform in precollege mathematics education. The current belief that school mathematics is a static and unchanging "discipline" is steeped in traditional values as well as in what has been taken to be mathematical reality. Choosing to focus on mathematics as a human activity does not change the fact that teachers must make decisions about what mathematics might be learned by their students. To the contrary, viewing school mathematics as not being a priori involves more as well as deeper decisions than when taking it as being given prior to teaching. When faced with the responsibility to make decisions about what might be learned, teachers can become mathematically active and seriously reconstruct the mathematical knowledge they only thought they knew. In that reconstructive process, they can establish an encompassing network of mathematical concepts and operations that deepens, unifies, and extends their conceptions of what might be learned. This network provides professional mathematics teachers with a sense of ownership of what they teach and with the power and independence to rise above the current compartmentalization found in school mathematics textbooks. By taking command of the mathematics that should be worked on with their students, they can opt to stress a network of mathematical relations they know is appropriate for their students. No one else is in a better position to make these decisions, and such independence can serve only to improve the mathematical knowledge of students.

The generative power of mathematics students is impressive when they are in learning environments that are conducive to constructive activity. However, this generative power is essentially uncharted. This makes adaptive mathematics teaching especially exciting because the involved teachers are in the best position to draw the maps. In the process, the mathematical knowledge of the students becomes a part of the knowledge of the teachers, and comments like "You just don't understand!" can be countered. It is especially important for the students to learn mathematics in such a way that it helps them organize their experiential world. Then questions like "Why do I have to learn this junk? I'll never use it" can be answered by the students themselves. Students must gain the confidence that they can engage in productive mathematical reasoning in such a way that it helps them to solve the problems they encounter and to see problems that would not exist for them had they not learned mathematics.

REFERENCES

Born, Max. *My Life and My Views.* New York: Charles Scribner's Sons, 1965.

Fischbein, Efraim, Maria Deri, Maria S. Nello, and Maria S. Marino. "The Role of Implicit Models in Solving Verbal Problems in Multiplication and Division." *Journal for Research in Mathematics Education* 16 (January 1985): 3–17.

Inhelder, Bärbel, and Jean Piaget. *The Growth of Logical Thinking from Childhood to Adolescence.* London: Routledge & Kegan Paul, 1958.

National Council of Teachers of Mathematics, Commission on Standards for School Mathematics. *Curriculum and Evaluation Standards for School Mathematics.* Reston, Va.: The Council, 1989.

National Research Council. *Everybody Counts: A Report to the Nation on the Future of Mathematics Education.* Washington, D.C.: National Academy of Sciences, 1989.

Steen, Lynn A. "Forces for Change in the Mathematics Curriculum." Paper presented at The School Mathematics Curriculum: Raising National Expectations, University of California at Los Angeles, 1986.

Steffe, Leslie P. "Children's Construction of Number Sequences and Multiplying Schemes." In *Number Concepts and Operations in the Middle Grades,* edited by James Heibert and Marlyn Behr, pp. 119–40. Reston, Va.: National Council of Teachers of Mathematics; Hillsdale, N.J.: Lawrence Erlbaum Associates, 1988.

Treffers, Adrian. *Three Dimensions: A Model of Goal and Theory Description in Mathematics Instruction—the Wiskobas Project.* Boston: D. Reidel Publishing Co., 1987.

von Glasersfeld, Ernst. "Communication and Environment." In *Transforming Early Childhood Mathematics Education,* edited by Leslie P. Steffe and Terry Wood. Hillsdale, N.J.: Lawrence Erlbaum Associates, forthcoming.

Wertheimer, Max. *Productive Thinking.* New York: Harper & Row, Publishers, 1959.

Wood, Terry, and Erna Yackel. "The Development of Collaborative Dialogue within Small Group Interaction." In *Transforming Early Childhood Mathematics Education,* edited by Leslie P. Steffe and Terry Wood. Hillsdale, N.J.: Lawrence Erlbaum Associates, 1989.

6

Small-Group Cooperative Learning in Mathematics

Neil Davidson

SMALL-group cooperative learning can be used to foster effective mathematical communication, problem solving, logical reasoning, and the making of mathematical connections—all key elements of NCTM's *Curriculum and Evaluation Standards for School Mathematics* (NCTM 1989). Small-group cooperative learning methods can be applied with all age levels of students, all levels of the mathematics curriculum from elementary school through graduate school, and all major topic areas in mathematics. Moreover, small groups working cooperatively can be used for many different instructional purposes: in the discussion of concepts, inquiry/discovery (often using manipulative materials), problem solving, problem posing, proofs of theorems, mathematical modeling, the practice of skills, review, brainstorming, sharing data from different groups, and the use of technology.

This article will address many of the questions most frequently asked by teachers about cooperative learning: (1) What is the rationale for small-group cooperative learning in mathematics? (2) What are the outcomes of cooperative small-group learning in mathematics? (3) What are appropriate leadership styles for the teacher? (4) How does one foster cooperative behavior among students? (5) How are groups formed? (6) How are students held accountable and graded? (7) How frequently do group activities occur? (8) What types of mathematical activities are most appropriate for small-group learning? (9) What resource materials for group work are available to the teacher? (10) What do students and teachers perceive as the strengths and limitations of cooperative learning in mathematics? (11) How does a teacher begin group work in mathematics?

RATIONALE

Why does cooperative learning have a place in mathematics instruction? The learning of mathematics is often viewed as an isolated, individualistic, or competitive matter—one sits alone and struggles to understand the material or solve the assigned problems. This process can often be lonely and frustrating. Perhaps it is not surprising that many students and adults are afraid of mathematics and develop "math avoidance" or "math anxiety." They often believe that only a few talented individuals can function successfully in the mathematical realm.

Small-group cooperative learning addresses these problems in several ways:

- Small groups provide a social support mechanism for the learning of mathematics. "Small groups provide a forum in which students ask questions, discuss ideas, make mistakes, learn to listen to others' ideas, offer constructive criticism, and summarize their discoveries in writing" (NCTM 1989, p. 79).
- Small-group learning offers opportunities for success for all students in mathematics (and in general). The group interaction is designed to help all members learn the concepts and problem-solving strategies.
- Mathematics problems are ideally suited for group discussion because they have solutions that can be objectively demonstrated. Students can persuade one another by the logic of their arguments.
- Mathematics problems can often be solved by several different approaches, and students in groups can discuss the merits of different proposed solutions.
- Students in groups can help one another master basic facts and necessary computational procedures in the context of games, puzzles, or the discussion of meaningful problems.
- The field of mathematics is filled with exciting and challenging ideas that merit discussion. One learns by talking, listening, explaining, and thinking with others, as well as by oneself.
- Mathematics offers many opportunities for creative thinking, for exploring open-ended situations, for making conjectures and testing them with data, for posing intriguing problems, and for solving nonroutine problems. Students in groups can often handle challenging situations that are well beyond the capabilities of individuals at that developmental stage.

OUTCOMES

The outcomes of cooperative learning methods have generally been quite favorable. General reviews of research have been presented by Sharan, Slavin, and the Johnsons; these are described in Slavin et al. (1985). Reviews by Davidson (1985), Davidson and Dees (forthcoming), and Webb (1985) focus specifically on mathematics learning; these first two mathematics reviews address achievement and the third addresses group interaction.

Research has shown positive effects of cooperative learning in the following areas: academic achievement, self-esteem or self-confidence as a learner, intergroup relations including cross-race friendships, social acceptance of mainstreamed children, and ability to use social skills (if these are taught).

Davidson (1985) and Davidson and Dees (forthcoming) reviewed about eighty studies in mathematics comparing student achievement in cooperative learning versus whole-class traditional instruction. In over 40 percent of these studies, students in the small-group approaches significantly outscored the control students on individual mathematical performance measures. In only two studies did the control students perform better, and both of these studies had irregularities in design.

The effects of cooperative learning of mathematical skills were consistently positive when there was a combination of individual accountability and team recognition for commendable achievement. The effects of small-group learning were nonnegative (i.e., not significantly different from traditional instruction) if the teacher had no prior experience in small-group learning, was not aware of well-established methods, and did very little to foster group cooperation or interdependence.

For many teachers the social benefits of cooperative learning are at least as important as the academic effects. Cooperative learning is a powerful tool for increasing self-confidence as a learner and for fostering true integration among diverse student populations.

CLASSROOM PROCEDURES

Cooperative learning involves more than just putting students together in small groups and giving them a task. It also involves giving very careful thought and attention to various aspects of the group process.

A class period might begin with a meeting of the entire class for an overall perspective. This meeting may include the presentation of new material, class discussion, posing problems or questions for investigation, and clarifying directions for the group activities.

The class is then divided into small groups, usually with four members apiece. Each group has its own working space, which might include a flip chart or a section of the chalkboard. Students work together cooperatively in each group to discuss mathematical ideas, solve problems, look for patterns and relationships in sets of data, make and test conjectures, and so on. Students actively exchange ideas and help each other learn the material. The teacher takes an active role, circulating from group to group, giving assistance and encouragement, and asking thought-provoking questions as needed.

The Teacher's Role

In each type of small-group learning, a number of leadership and management functions must be performed. Although some of them may be explicitly delegated to the students, these functions are generally handled by the teacher and include the following:

- Initiating group work
- Presenting guidelines for small-group operation
- Fostering group norms of cooperation and mutual helpfulness
- Forming groups
- Preparing and introducing new material
- Interacting with small groups
- Tying ideas together
- Making assignments of homework or in-class work
- Evaluating student performance

Each of these functions can be performed in different ways and to varying degrees, depending on the model of small-group instruction in effect.

The behavior of the teacher will vary for different phases of instruction, such as introducing new material, facilitating group activities, and summarizing. A teacher with small groups can introduce new material and pose problems and questions for discussion or investigation (*a*) orally, with a whole-class discussion at the beginning of a period or with individual groups at appropriate moments; or (*b*) in written form, with teacher-made worksheets or special texts designed for small-group learning. In any event it is essential to make sure that students understand the mathematical problem situation. (Several examples of activities are given later.)

The teacher provides guidance and support during small-group activities, observing the group interaction and their solutions on the board, and, while visiting particular groups, checking their solutions, giving hints, clarifying notation, making corrections, answering some questions, and so on. The teacher also performs social functions, such as providing encouragement, drawing members into the discussion, and helping the groups function more cooperatively. The teacher should behave in a friendly and constructive manner and strike a balance between giving too much and too little assistance. It may take a while for the teacher to become adept at observing, diagnosing difficulties, and intervening in a facilitative way.

In summarizing discussions with the whole class, the teacher may need to answer certain questions, serve as discussion moderator, and clarify and summarize what the students have found. An overall synthesis by the teacher is needed from time to time, since students in the groups sometimes "see the trees but lose sight of the forest." Occasionally it is useful to hear brief summary statements from different small groups. This is especially powerful when different groups have tackled different aspects of a complex situation, perhaps leading to a generalization. However, end-of-period summaries are not always necessary and should not become a ritual.

Fostering Cooperation

To help the students learn how to cooperate with one another, the teacher might present a set of guidelines for group behavior:

1. Work together in groups of four.
2. *Cooperate* with other group members.
3. Achieve a group solution for each problem
4. Make sure that everyone understands the solution before the group goes on.
5. Listen carefully to others and try, whenever possible, to build on their ideas.
6. Share the leadership of the group.
7. Make sure that everyone participates and no one dominates.
8. Take turns writing problem solutions on the board.
9. Proceed at a pace that is comfortable for your own group.

Different sets of guidelines are possible. Difficulties in group interaction can usually be analyzed and cleared up by reminding students about the guidelines, whichever ones are stated.

One way to improve group functioning is to reflect on the communication and interaction process occurring in each group. Each member of the group can address the following questions while other members listen: How are you contributing to the successful operation of this group? What can you do to make it function even better?

A more elaborate set of questions for discussion follows: (1) Did your group achieve at least one solution to the problem or task? (2) Did everybody understand the solution? (3) Did people ask questions when they didn't understand? (4) Did people give clear explanations? (5) Did everyone have a chance to contribute ideas? (6) Did people listen to one another? (7) Did any one person take over the group? (8) Did the group really work together on the task? (9) Was there enough time for exploration?

Group Size and Formation

It is necessary to form work groups of small size, since the opportunity for active participation decreases as the group size increases. In mathematics classes, groups with four members seem to work best. They are large enough to generate ideas for the discussion and solution of challenging problems, yet not be decimated by the absence of one member. They are small enough to permit active participation, to allow clustering around a chalkboard panel, and not to require a leader or elaborate organizational structure. Groups of four can also split into pairs for occasional computational practice or simple application problems.

Generally, if one gives great care and attention to forming the groups, the groups will function better and there will be less need to switch groups frequently. Although groups can be formed by random assignment, heterogeneous grouping ensures a mixture of mathematical achievement, gender, and race/ethnicity. An occasional use of sociometric choice by students is possible, but homogeneous grouping is usually not recommended. Proponents of different schemes sometimes express strong ideological viewpoints about grouping procedures.

An experienced teacher can usually work comfortably with as many as six or seven groups but might feel overly extended with eight groups. In very large classes, the teacher may need an aide for help with group supervision; a more advanced student can often be an effective aide.

Evaluation

A variety of grading schemes are compatible with small-group instruction, including in-class tests and quizzes, take-home tests, homework, classwork (consider attendance, participation, and cooperation), self-evaluation, and peer evaluation. If a teacher gives tests on a specific date, that date should realistically allow all groups to have finished the material before the test date without rushing. If teachers give grades for classwork (including attendance, participation, cooperation) they should *not* grade individual mathematical performance during class; doing so will foster competition and destroy cooperation. Some teachers have found that evaluation measures such as group projects or the occasional use of group tests on which all members receive the same grade work well. Personal philosophy has a great bearing on these decisions.

Frequency of Use

The small-group method can be used as a total instructional system or in combination with other methods. Groups can be used all the time, on specific days of the week, during portions of any class period, or for specific topics. I personally prefer to use small groups for most of the class time, except in a few multisection departmentalized courses taught on a rigid time schedule at breakneck pace with uniform hour exams.

MATHEMATICAL ACTIVITIES

Let us now focus on curriculum-related issues. What constitutes a good problem for small-group discussion? Crabill (1990) suggests that a good problem for group exploration has the following attributes (pp. 215–16):

> (1) It is presented in a meaningful context and in everyday language whenever possible . . . . (2) It is easy to state. That is, the problem is clearly defined, even though it may not be easy to solve. (3) It is easy to visualize physically. That is, it is not abstract, even though it may lead to an abstract generalization later. (4) It stimulates student questions that may be better than the original question. This is the most important attribute. These attributes of a good problem are especially valuable when starting a new topic or starting new learning groups.

Two examples of effective problems for small groups follow.

Activity 1

Students in small groups are capable of "discovering" several classical summation formulas. It is best to start with a simple situation, such as the following:

Consider sums of consecutive odd integers, such as 1, 1 + 3, 1 + 3 + 5, 1 + 3 + 5 + 7, 1 + 3 + 5 + 7 + 9, ... Compute these sums and record the answers in this table:

Number of terms	1	2	3	4	5	6	7
Sum of odd integers	1	4					

State a formula (rule, generalization) that tells how to compute easily the sum of any number of consecutive odd integers (starting with 1). Use your rule to compute the sum of the first 1000 odd integers.

Now let's look for other formulas for the sums of consecutive—

- even integers: 2, 2 + 4, 2 + 4 + 6, 2 + 4 + 6 + 8, . . .
- integers: 1, 1 + 2, 1 + 2 + 3, 1 + 2 + 3 + 4, . . .
- integers cubed: 1^3, $1^3 + 2^3$, $1^3 + 2^3 + 3^3$, $1^3 + 2^3 + 3^3 + 4^3$, . . .

Compute these sums and record the answers in the following table:

Number of terms	1	2	3	4	5	6	7
Sum of even integers	2	6					
Sum of integers	1	3					
Sum of cubes	1	9					

Hint (if needed): Compare different lines in the table.

State three formulas (rules, generalizations) that tell how to compute easily (1) the sum of any number of consecutive even integers (starting with 2), (2) the sum of any number of consecutive integers (starting with 1), and (3) the sum of the cubes of any number of integers (starting with 1^3). Use your rules to compute each of these sums, with each having 1000 terms.

Comment: If students are familiar with the use of variables, they may write their formulas in symbolic notation:

$$1 + 3 + \cdots + (2n - 1) = n^2$$
$$2 + 4 + \cdots + 2n = n(n + 1)$$
$$1 + 2 + \cdots + n = n(n + 1)/2$$
$$1^3 + 2^3 + \cdots + n^3 = [n(n + 1)/2]^2$$

Otherwise, they may simply state their results in words.

Activity 2

Sometimes a situation can be divided into tasks of roughly equal difficulty. Each group explores one such task or aspect of the situation and then presents its results to the whole class. For example, in first introducing the concept of the slope of a line, the teacher can ask each group to make several graphs on the same axis by plotting points. The assignments can be as follows:

Group 1: Graph $y = x$, $y = 2x$, $y = 3x$
Group 2: Graph $y = -x$, $y = -2x$, $y = -3x$
Group 3: Graph $y = (1/2)x$, $y = (1/3)x$, $y = (1/4)x$
Group 4: Graph $y = (-1/2)x$, $y = (-1/3)x$, $y = (-1/4)x$

Each group then displays its graphs on the chalkboard or on large pieces of paper. These displays allow students to see the effects of changing the coefficient of x and permit generalizations regarding these effects on the slope of the line. Similar explorations can be conducted with intercepts.

One of the major issues in designing group activities is the amount of guidance (or structure) to give in the directions. Higher levels of guidance imply lower levels of open-ended exploration, inquiry, or discovery and perhaps less intellectual excitement and "thrill of discovery," but they also provide for more efficient use of time, more rapid coverage of material, and less student frustration. There are trade-offs in any decision about levels of guidance. Other variables to be considered are the amount of time available, the perceived ability of the students to handle challenges and possible frustration, and the importance of open-ended exploration for the particular topic at hand. It is often hard to gauge this in advance; classroom trials may be needed to redesign the activities according to the actual experiences of the groups. See Davidson, McKeen, and Eisenberg (1973) for further suggestions. A number of people have created mathematics curriculum resources for use in cooperative learning situations at different levels; see Davidson (1990) for detailed descriptions.

Perceptions of Teachers and Students

In attitude surveys given over a period of years, the main problems that teachers and students expressed about cooperative learning in mathematics were as follows: Concerns about covering enough material, initial difficulties in forming effective groups, barriers to fostering cooperation among students, occasional conflict or frustration with overly difficult mathematical problems, providing high-quality instructional materials, and handling a major shift in the roles of teacher and student. Although student attitudes toward this method of instruction are generally favorable, the degree to which they are favorable depends on the teacher's experience and skill in handling the problem areas mentioned above.

There are many advantages to learning mathematics in small, cooperative groups. The following positive points are frequently mentioned by teachers and students in responding to attitude surveys: Students are actively involved in learning mathematics while working at a comfortable pace. They learn to cooperate with others, to improve their social skills, and to communicate in the language of mathematics. The classroom atmosphere tends to be relaxed and informal, help is readily available, questions are freely asked and answered, and misconceptions become quickly apparent and are readily resolved. Students tend to become friends with their group members across traditional boundaries of race, ethnicity, or sex. The teacher-student relationship tends to be more relaxed, pleasant, and closer than in a traditional approach. Teachers benefit from some intellectual companionship with their students and often find themselves invigorated professionally and less prone to burnout. The usual disciplinary problems of talking and moving around are eliminated by definition. In addition, many students maintain a high level of interest in the mathematical activities. Students are not bored in class; many of them like mathematics more than when involved in teacher-centered approaches. Finally, students have an opportunity to pursue the more challenging and creative aspects of mathematics and to become more confident problem solvers while acquiring at least as much information and skill as when they are taught with more traditional approaches.

GETTING STARTED

Here are a few tips for teachers just getting started in implementing cooperative learning: (1) Begin with a simple approach such as think-pair-share. The teacher poses questions to the class, where students are sitting in pairs. Students *think* of a response individually for a given period of time, then *pair* with their partners to discuss the question and reach consensus. The teacher then asks students to *share* their agreed-on answers with the rest of the class. This method can be used during teacher presentations whenever students appear to be confused or not attentive. Students are asked to discuss the content and attempt to explain it to one another or else formulate a question for the class. The method can also be used for previously planned brief discussions or short practice activities. It has a powerful, immediate effect in livening up the class. (2) Look for opportunities within the regular curriculum to use groups. For example, class time that is normally devoted to individual seatwork can be changed into group work. The textbook may include a good selection of exercises and problems for this purpose. (3) Give very clear, step-by-step directions and check to make sure that students understand them. (4) Start with a class that you think will respond favorably to cooperative learning. (5) Don't feel that you must establish very tight control for weeks before beginning group activities. Group work can become part of your management system. (6) Clearly inform the principal or building administrator and parents what you are doing and why you are doing it. (7) If at all possible, find a colleague who will use similar methods. Two teachers together can provide strong mutual support. (8) Expect that group activities will not necessarily go smoothly at first; it usually takes two or three weeks for students to begin functioning well in groups. (9) Remember that change is a gradual process, not an event. Don't try to change everything all at once.

REFERENCES

Crabill, Calvin. "Small-Group Learning in the Secondary Mathematics Classroom." In *Cooperative Learning in Mathematics: A Handbook for Teachers,* edited by Neil Davidson, pp. 215–16. Reading, Mass.: Addison-Wesley Publishing Co., 1990.

Davidson, Neil. *Cooperative Learning in Mathematics: A Handbook for Teachers.* Reading, Mass.: Addison-Wesley Publishing Co., 1990.

________. "Small Group Cooperative Learning in Mathematics: A Review of the Research." In *Cooperative Learning Research in Mathematics,* edited by Neil Davidson and Roberta Dees. *Journal for Research in Mathematics Education* Monograph Series. Reston, Va.: National Council of Teachers of Mathematics, forthcoming.

________. "Small Group Learning and Teaching in Mathematics: A Selective Review of the Research." In *Learning to Cooperate, Cooperating to Learn,* edited by Robert Slavin, et al., pp. 211–30. New York: Plenum Press, 1985.

Davidson, Neil, Ronald McKeen, and Theodore Eisenberg. "Curriculum Construction with Student Input." *Mathematics Teacher* 66 (March 1973): 271–75.

National Council of Teachers of Mathematics, Commission on Standards for School Mathematics. *Curriculum and Evaluation Standards for School Mathematics.* Reston, Va.: The Council, 1989.

Slavin, Robert, et al., eds. *Learning to Cooperate, Cooperating to Learn.* New York: Plenum Press, 1985.

Webb, Noreen. "Verbal Interaction and Learning in Peer-directed Groups." *Theory into Practice* 24 (1985): 32–39.

7

An Eclectic Model for Teaching Elementary School Mathematics

Thomas E. Rowan
Nancy D. Cetorelli

IN RECENT years many writers and researchers have pointed out the need for improving the methods and content of mathematics teaching at the elementary school level (McKnight et al. 1987, Gross 1988, Dossey et al. 1988). At the same time, they have advocated various approaches to teaching mathematics, including the following: direct instruction, cooperative learning groups, directed discovery (with the teacher as facilitator), Teacher Expectation and Student Achievement (TESA), and mastery learning. We recommend that elementary school teachers know and use a variety of approaches as part of a carefully developed instructional plan.

Taking such an eclectic approach can benefit student learning in several ways. The first is that it enables teachers to identify and use their personal strengths while balancing those strengths in an effective way with less familiar or more difficult instructional approaches. Teachers can use the approach with which they are most comfortable more frequently but can develop enough facility with other approaches to use them at regular intervals. A second benefit of an eclectic approach is that it motivates students. Changes in routine help foster student interest and involvement and promote a higher degree of original learning (Hunter 1982, p. 111). A third and more subtle benefit of such an eclectic approach is the opportunity to match the nature of the mathematics content to the method of instruction. When we think through such matching possibilities, we gain the secondary benefit of enhancing our own understanding of the mathematics content being considered.

Teachers who systematically monitor their students' progress and use this information to carefully plan lessons with a variety of teaching approaches can avail themselves of the aforementioned benefits. The important thing is to plan for, and build in, the variation. Increased student learning should result from the variety. Careful selection of methods to match the content and the needs of the students will further enhance the overall result. The gist of our recommendation, then, is *do not* use the same approach to instruction day in and day out but *do* know and use a variety of mathematics teaching approaches in a carefully planned program of instruction.

Planning is the key to assuring that instruction will be as effective as possible. Both long-range and short-range planning should be done (see fig. 7.1 for a suggested long-range planning form). A long-range plan assures comprehensive coverage of the content and allows the teacher to take advantage of opportunities that relate to topics yet to come, such as an effective teaching suggestion found in a professional journal. Short-range planning specifies the methods and materials to be used and assures that all the materials are on hand. Short-range plans also allow for adjusting the long-range schedule to the needs of the class or group. However, teachers should minimize as much as possible on-the-spot changes in instructional activities, since even if all the necessary mathematical information is known, the important manipulative and other instructional materials so essential to the teaching of concepts would not likely be at hand.

PLANNING AND MANAGING EFFECTIVE MATHEMATICS INSTRUCTION

An eclectic instructional approach uses a variety of activities in the context of a daily routine with a regular pattern. This combination contributes to good class management but still allows for variation within the routine. A recommended daily routine is discussed below:

1. Warm-up activities
2. Systematically presented lessons
3. Announced or unannounced review
4. Closing activities

Warm-up Activities

Warm-up activities are short activities used at the beginning of the class period to get the students as quickly as possible into a frame of mind for doing mathematics. Some examples of warm-up activities follow:

- Reading a brief historical story of mathematics and discussing it (e.g., the story of the invention of the metric system—could be continued across several days to maintain brevity)

Long-Range Plan—Mathematics
(By Categories)

Teacher ______________________

Levels ______________________

Month	Sept.	Oct.	Nov.	Dec.	Jan.	Feb.	March	April	May	June
Instruction										
Review										
Grading Period (9 wk)	1st Grading Period		2d Grading Period			3d Grading Period		4th Grading Period		
Holidays										

Fig. 7.1

- Solving an amazing problem (e.g., the height of a million pennies)
- Playing buzz (e.g., saying "Buzz" for multiples of 3 when counting)
- Giving a five-problem quiz on yesterday's lesson
- Having a short basic-facts practice activity
- Presenting a "Problem of the Week"
- Solving a mathematics riddle (e.g., "I am a polygon with fewer than five sides and more than three angles; who can I be?")
- Writing a paragraph about yesterday's lesson (e.g., What do you remember best? What was easiest? What was hardest? Write a sample problem for your classmates. How would you teach the lesson to your younger brother or sister?)

Systematically Presented Lessons

Brief descriptions of various instructional approaches are given below, not as detailed descriptions or a complete list, but only to indicate the variety of an eclectic classroom:

- *Direct instruction*. The teacher takes a strong leadership role and uses a particular sequence of events during each lesson. Included are daily review, a lesson or demonstration by the teacher, supervised practice, and independent practice. The teacher is actively in charge of everything that happens in the classroom. The direct lesson or demonstration part of this scheme would be a model of one approach to a systematically presented lesson in the eclectic classroom. For further details on this instructional technique, see Good, Grouws, and Ebmeier (1983) and the article by Kanold in this yearbook.
- *Small cooperative groups*. The students are formed into groups in which they work together to attain the skills and concepts that are the object of the instruction. Various methods are used to define the groups; they are not self-selected. Planned selection enables the students to work efficiently toward the learning goals. Just putting students haphazardly into groups will not produce a magical change. It is what is done in those groups that is important. Students should be trained in the requisite interactive skills so that group members can work together effectively. Students who choose their own groups and are allowed to make their own rules may not focus well on the objectives and often lose opportunities for social development. Examples of well-organized approaches to small cooperative groups can be found in Johnson and Johnson (1984, pp. 25–41), Slavin (1984) and the article by Davidson in this volume.
- *Discovery*. This approach may be used either as a separate approach to instruction or in the context of cooperative groups. Its main characteristic

is that children (together or independently) manipulate materials or information that will lead them to discover the desired mathematical principle. The occasional truly original discovery can happen in a less structured environment. However, a lesson with no objective can leave both teacher and students frustrated and drifting. An example of a discovery lesson is to have students explore the partitioning of base-ten blocks to discover a meaning and process for division (Burns et al. 1988).

• *Discussion.* A discussion lesson is often productive, but every lesson should not revert to this method. The question-and-answer method can be planned for individual, small-group, or large-group organizations, but it may be more effective with individuals and small groups. In a large group care should be taken that too many students are not passive while one student responds. Discussions with larger groups can be made more effective by employing methods to assure active participation by all students, such as monitoring individual responses (e.g., small slates or numeral cards held up by each student) or "pairing and sharing." Questions used with groups can also be more effective if they are formulated to yield process rather than product answers. For example, instead of just being asked to name and identify triangles or to memorize rules for finding the areas of polygons, students could be asked to show how polygons can be partitioned into rectangles or triangles as a vehicle for finding the areas of those polygons. It is important to keep in mind that learning objectives (and student readiness) should determine the selection of instructional strategies, materials, and learning activities, not the other way around.

Announced or Unannounced Review

Review should be a regular part of the program and should be distributed at thoughtful intervals. It has long been known to be effective for improving learning and retention, but it seems not to be used as often as it should be. The reason may be the difficulty of working such review into the regular flow of the lesson. Careful and flexible long-range plans as previously discussed are essential if review is to occur on a regular and effective schedule and to incorporate the idea of variety. Some approaches that can be used for review follow:

• *Announced reviews.* Have a specific time each week for review of some sort. This period could even be given a name, such as "Remember When" time. Although the time would be known, the topic to be reviewed and the method of review may or may not be known. Most of these reviews should be short, perhaps ten minutes or so. Occasionally, they can be extended to a whole class period for particularly important topics. Review can also be an expected, intergral part of regular homework assignments.

• *Unannounced reviews.* Some teachers prefer to have unscheduled reviews. Such reviews can take many forms: a warm-up activity, a quick quiz, part of a quiz or test on current studies, an activity with manipulative materials, a small-group buzz session, writing a letter to a friend far away to explain the mathematics being reviewed.

Closing Activities

It is important for students to leave a lesson with an upbeat feeling rather than feel that it died from a lack of interest. A good closing activity helps to foster a good feeling, even when the lesson is on a routine topic. The closing should be motivational and not present additional instruction or review. A review of important points of the lesson and any assignment of homework should precede the closing. The previously mentioned game of buzz or an amusing problem can be used as a closing activity, or one of the following:

• *Quick puzzle.* This kind of activity could be something as simple as having students act as detectives to solve clues for "Sherlock Math." The clues might be "What are the numbers that meet all these requirements: (*a*) they are the sum of two odd numbers, (*b*) they are divisible by 6, (*c*) they are less than 50, and (*d*) their digits sum to 6?" The students, individually or in groups, would write their answers on a slip of paper and hand them to the "Sherlock" of the day, who might be the student who invented the puzzle. Sherlock checks and reports the results the next day, either individually or as group data. Simple rewards accessible to everyone (e.g., stickers) can help to motivate the students.

• *Journal writing.* This activitiy gives students an opportunity to ask questions about the topic of the day, to reflect quietly on the lesson, to point out what they feel is the most important part of that topic, or simply to express freely whatever is on their minds about the content or method of the lesson. It gives the teacher an opportunity to gather feedback that may be helpful in the planning process. Keeping a journal also provides an opportunity to interrelate mathematics and writing. Science or social studies could be worked in by having the student write a paragraph that tells how the day's mathematics, science, and social studies lessons all relate to one another. Sketches of important ideas might be acceptable entries on some days for students who are more comfortable with drawing. Further information on the instructional uses and potential of journal writing appears in the article by Azzolino later in this book.

• *Game of "Who Has?"* This game can provide practice or review in a motivational setting. It is played by making a set of index cards, each with a question and the answer to a question from another of the cards. The cards are distributed to the students, one of whom begins by reading a question on his or her card. Whoever has the answer to that question on a card reads it and then reads the question on that card. The game continues until the initial question reappears. Each student should have one or more cards.

SUMMARY AND CONCLUSION

This article promotes the use of an eclectic approach to the teaching of elementary school mathematics. The recommended eclectic approach is planned, well structured, and matched to the concepts and content that underlie the learning objectives of each lesson. The importance of long-range planning is emphasized by the inclusion of a prototype form for yearlong planning (fig. 7.1). The approach provides a structure for short-range plans (daily lesson design), including warm-up activities, systematically presented instruction, announced or unannounced review, and closing activities.

Variety that supports the critical outcomes of increased student involvment and learning is the basic premise of the eclectic approach. It has been said that if the only tool one has is a hammer, then every problem tends to be treated as a nail. Only the classroom teacher who knows several effective instructional practices can use them to elicit these outcomes from students. This method of improving instruction can be used by any teacher, although it becomes more effective when supported by a good mathematics background.

REFERENCES

Burns, Marilyn, and Bonnie Tank. *A Collection of Math Lessons from Grades 1–3*. Los Angeles: Math Solution, 1988.

Dossey, John A., Ina V. S. Mullis, Mary M. Lindquist, and Donald L. Chambers. *The Mathematics Report Card: Are We Measuring Up?* Princeton, N.J.: Educational Testing Service, 1988.

Gross, Susan. *Participation and Performance of Women and Minorities in Mathematics*. NSF Study. Rockville, Md.: Montgomery County Public Schools, 1988.

Good, Thomas, Douglas Grouws, and H. Ebmeier. *Active Mathematics Teaching*. New York: Longman Publishing, 1983.

Hunter, Madeline. *Mastery Teaching*. El Segundo, Calif.: TIP Publications, 1982.

Johnson, David W., et al. *Circles of Learning: Cooperation in the Classroom*. Alexandria, Va.: Association for Supervision and Curriculum Development, 1984.

McKnight, Curtis, F. Joe Crosswhite, John A. Dossey, Edward Kifer, Jane O. Swafford, Kenneth J. Travers, and Thomas J. Cooney. *The Underachieving Curriculum: Assessing U.S. School Mathematics from an International Perspective*. Champaign, Ill.: International Association for the Evaluation of Educational Achievement, Stipes Publishing Co., 1987.

Slavin, Robert. *Learning to Cooperate, Cooperating to Learn*. New York: Plenum Publishing Co., 1984.

8

Mathematics as Communication: Using a Language-Experience Approach in the Elementary Grades

Frances R. Curcio

THE language of mathematics is highly dependent on its unique symbols and technical terminology. If children are to succeed in their study of mathematics, they must be allowed to explore, examine, and express the mathematical relationships they experience in everyday life rather than be required to memorize meaningless formal definitions and notation (Curcio 1985; Skemp 1982).

One of the general themes of the *Curriculum and Evaluation Standards for School Mathematics* focuses on mathematics as communication (National Council of Teachers of Mathematics 1989). Communicating mathematical ideas should be related to children's experiences at different grade levels. Listening, speaking, reading, and writing help children to clarify their ideas and share them with others. A natural way to help children develop the language of mathematics is by employing a language-experience approach to the teaching of mathematics.

Most elementary school teachers are not strangers to the language-experience approach as a method of teaching beginning and remedial reading (Harris and Sipay 1980). This personalized, reality-based approach encompasses such activities as listening, speaking, reading, and writing, where children are guided to express their reactions, ideas, and feelings regarding situations shared by the children in the classroom or on a field trip. Large-

The author would like to express thanks to the teachers and children at P.S. 104K, Brooklyn, New York, for their interest and participation in the activities described in this article.

and small-group activities that highlight this approach are also successful in teaching mathematics because they help make learning meaningful. In this article, the language-experience approach to the teaching of elementary school mathematics will be discussed and illustrated using fraction concepts.

DEVELOPING FRACTION CONCEPTS

Results of the most recent National Assessment of Educational Progress (Kouba et al. 1988) indicate that children have a limited understanding of fraction concepts. Using abstract symbols, terminology, and forms of representation without developing meaning based on children's experiences and readiness may cause some of the difficulties that they exhibit. Instead of dealing with one numeral at a time, as they do with the natural numbers, children are faced with two numerals arranged vertically separated by a vinculum. And if this new configuration were not enough to cause problems, add to it the "new" words, *fraction* and *ratio*. Although children may be familiar with such words as *half, third,* and *quarter,* their exposure to fractional parts beyond these three is limited (Kieren 1980). Building on children's informal language may help enhance their understanding of fractions (Kieren 1984).

Since there are several interpretations and uses of fractions (e.g., the parts-of-a-whole meaning, the quotient meaning, and the ratio meaning), becoming familiar with the appropriate interpretation in a contrived classroom situation might cause confusion (Lay 1982). If children encounter the concept of fractions in their everyday life situations and learn to deal with them informally, why can't they experience success in the mathematics classroom? Understanding may be facilitated when teachers not only build on children's informal language but also use appropriate materials to represent the concepts to be developed. The underlying mathematical structure represented by manipulatives or by various situations must be brought to bear by allowing children to share their ideas and interpretations through discussing, writing, and reading about fractions.

Some examples of incorporating the parts-of-a-whole, quotient, and ratio interpretations of fractions within a language-experience framework of mathematics lessons are presented in the following sections. At each grade level the activities can be conducted in either large- or small-group settings. While some children are sharing their ideas, others should be listening and thinking of ideas to contribute or questions to ask. In the early grades the teacher should record the students' story on the chalkboard or overhead transparency for everyone to see. Beginning in grade 3, children should write stories based on their own experience with the activity. Stories should be read aloud and discussed.

Grades K–2: Exemplifying the Parts-of-a-Whole Meaning of Fraction

Ask the children, "Who has ever shared a fruit, cake, or candy bar with a friend? Tell us about it." The teacher should guide the discussion to focus on size, shape, and amount as well as feelings about sharing different-sized pieces. The emphasis should be on counting the number of same-sized (or equal) pieces in the whole. If children did not use it themselves, the teacher might want to introduce the term *one-half.* During the discussion the teacher should ask the children, "What is the difference between cutting the piece of fruit into two pieces and cutting it in half?" This idea should be demonstrated using real fruit. An example of a first-grade language-experience story follows:

> Ms. Curcio came to class 1-4. She put on an apron. She took out a green apple. She called two children up to share the apple. We told her to cut the apple into two pieces. One piece was small and one was big. This was not a good way to share. We had to cut the apple in two pieces the same size. Out of her pocket she took a golden apple. Afif told her to cut it in the middle. By cutting it in the middle, we get two pieces that are the same size. We call one piece *one-half.*

The teacher should read the story aloud, pointing to each word and then encouraging children to read along during subsequent readings. Children could, depending on their ability, read collectively and individually. Reading is facilitated because the story is recorded in their own language. Again depending on the children's ability, they may be expected to copy the experience story into their writing books and refer to it at a later time.

For a follow-up lesson, the children should be grouped in twos and fours with each group having one large apple. The task for each group is to describe to an adult how to cut the apple so that each child in the group will get a piece that is the same size. Thus some apples should be cut into two same-sized pieces and some should be cut into four same-sized pieces, depending on the number of children in each group. The children should tell another story based on what happened during the apple-sharing experience.

Some children in a group of two may use such language as "Cut it in the middle" or "Cut it into two pieces." Ideas that could be discussed include "One whole is more than one-half" and "One-half is less than one whole." With apples that are approximately the same size, use whole apples and portions of apples to illustrate these ideas. The teacher may want to point out and discuss the problem of asking for the "larger half."

Children in a group of four might say, "Cut it in the middle and then in the middle again," or just "Cut it in fours." The teacher may want to introduce such phrases as *one-quarter* and *one-fourth* if the children do not mention them. (If this is done, the teacher should be careful to relate these terms to situations with which the children are familiar; e.g., there are four quarters in one dollar, etc.) The emphasis should be on counting the number of equal pieces in the whole. It is appropriate to elicit from the children the idea that the size of the pieces each child in a particular group gets depends on the number of children in the group as well as the size of the apple before cutting it. Compare the part of the whole that each child in a two-member group receives with the part of the whole that each child in a four-member group receives. Children should compare one-fourth, one-half, and the whole apple. An example of a first-grade experience story follows:

> First we cut an apple into halves because we had two children to share an apple. Then Ms. Curcio called up two more children. We needed to cut the two pieces again. Joseph said to cut each piece in the middle. Now we have four same-sized pieces in one apple. These pieces are called *fourths*.

After focusing attention on any new words that might have been introduced, the teacher should have children copy the story in their writing books. The teacher should then allow children to discuss other similar situations that require cutting fruit, cake, or a pie (or some other whole) into equal portions.

Grades 3–4: Exemplifying the Quotient Meaning of Fraction

Ask the children, "How can three children share six oranges?" (Elicit the idea that six oranges to be shared among three children, or $6 \div 3$, would allow each child to have two oranges.) "How can three children share three oranges?" (Elicit the idea that three oranges to be shared among three children, or $3 \div 3$, would allow each child to have one orange.) "How can three children share two oranges?" (If children are not familiar with any of these situations, be prepared to have small groups of children act them out.)

Children should observe that to share the two oranges equally among three people, the oranges must be cut. "If there were only one orange, how could the three children share?" (Cut it into three equal pieces; i.e., make thirds). "What should be done with the other orange?" (Cut it into thirds, also.) Have children observe that in each situation, each child will receive less than one whole orange. Altogether, each child will receive two-thirds of the oranges (eliciting the idea that two oranges shared among three people, or $2 \div 3$, would allow each child to have 2/3 of the oranges.) Krissy's "experience story" highlights the activity:

> Today we did math. Ms. Curcio took two oranges out of her bag. Tina, Farayi, and Joseph came to the front of our room. We tried to figure

out how to *share* the two oranges with the three children. We cut the oranges into *thirds*. Each child got two-thirds of the oranges. We shared the oranges fairly. Then we shared two granola bars among three children. Each child got two-thirds of the granola bars. We can write $2 \div 3 = 2/3$."

After the children have the opportunity to write a story, read it, and discuss it, they should identify other items (e.g., food) that can be shared. For example, the third graders mentioned pears, bananas, pizza, apples, sandwiches, slices of cheese, and candy bars. The children were then arranged in groups of two, four, five, six, and eight. Each group picked an item of food. Depending on the number of children in each group, the amount of food they were allowed was less than the number in their group (e.g., two children had to share one candy bar; four children had to share three slices of cheese; six children had to share four bananas, etc.). The groups were given strips of paper on which to draw and color their assigned food items. Then they had to decide how to share the items among the members of their groups. Some ideas were as follows:

- For two children to share one candy bar, or $1 \div 2$, each child gets 1/2.
- If four children share three slices of cheese, or $3 \div 4$, each child gets 3/4.
- When six children share four bananas, or $4 \div 6$, each child gets 4/6.

It should be noted that during this activity, children are not being introduced to, and bombarded with, many new terms such as *numerator* or *denominator*. It is more important to present situations that highlight the use and the meaning of these terms than to have children repeat and memorize them without understanding (Curcio 1985). Also, notation should not be introduced until the children demonstrate basic understanding of the concept. It is better to prolong the use of spoken language than to rush children into using short-hand technical notation (Skemp 1982).

Grades 5–6: Exemplifying the Ratio Meaning of Fraction

Ask, "How many of you have ever gone to the supermarket to buy ice cream? How are the ice-cream bars packed?" Give them the opportunity to discuss different brands of ice cream and their preferences. (A previous night's assignment could be to check prices, bring in newspaper advertisements, or bring in empty ice-cream containers.)

Some brands have six bars in a package, and others have eight. The teacher should guide the discussion to highlight the unit cost to illustrate the ratio meaning of fraction (e.g., "In one store six bars of ice cream cost

\$3.00. In another store eight bars cost \$4.80. Which is the better buy?") Students could express the relationship as \$3.00 is to 6 bars as "how much" is to 1 bar (i.e., $\$3.00/6 = x/1$)? Similarly, \$4.80 is to 8 bars as "how much" is to 1 bar (i.e., $\$4.80/8 = y/1$)? At first, select numbers so that the computation can be done mentally and so that students can focus on the concept rather than on the computation. In examples selected from newspaper advertisements, students should be encouraged to use estimation strategies or use a calculator to determine the better buy. After discussing several examples, students should write an experience story to be shared with the class. Important ideas should be summaried. The following one written by sixth graders Jennifer, Patrick, Jason, Lisa, and Eddie is a good example:

> Today we had a math lesson on comparative shopping and unit cost. We learned that sometimes we can get more for a cheaper price and save money. We can use ratios or fractions to compare the cost of different amounts to get the better buy. A ratio is used to compare two numbers. We were able to use calculators to find the unit cost. The unit cost is the cost of one item. We can break down the price and the quantity to its least form. Then when we go to the store, we know which one to buy so we don't get ripped off. We helped each other and had a lot of fun.

For a follow-up lesson, arrange the class into groups of three or four students. Give each group a situation such as this one: "A 10-oz. package of muffins costs \$1.89. A 12-oz. package of muffins costs \$2.52. Which is the better buy? Why?" Also discuss the other factors to be considered before buying the muffins, such as the location of the stores, the perceived quality or reputation of each brand, and the ingredients. Each group should keep a record of what members say about the situation and how it is resolved. A reporter from each group should share comments with the rest of the class.

Finally, each group should collect, or be supplied with, advertisements and write a story problem about comparative shopping. The problems can be traded among the groups. Students should discuss and solve them. A story problem written by Patrick, Chris, and Meropi follows:

> In K.K. Supermarket, a 20-oz. package of frozen pizza costs \$1.99. In Met Supermarket, a 24-oz. package of frozen pizza costs \$2.69. Which is the better buy? Why?

CONCLUSION

By using a language-experience approach, children have an opportunity to use everyday language to express ideas about mathematical concepts presented in the context of an everyday life situation, clarify their thinking, and reflect on the mathematical ideas they encounter. In accordance with the NCTM's *Curriculum and Evaluation Standards,* the language-experience approach can help bridge the gap that exists between mathematics in the real world and mathematics in school. This gap becomes increasingly pronounced in schools serving culturally diverse populations. Special techniques for dealing with these populations are illustrated in the next section of this yearbook.

REFERENCES

Curcio, Frances R.. "Making the Language of Mathematics Meaningful." *Curriculum Review* 24 (March/April 1985): 57–60.

Harris, Albert J., and Edward R. Sipay. *How to Increase Reading Ability.* 7th ed. New York: Longman, 1980.

Kieren, Thomas E. "One Point of View: Helping Children Understand Rational Numbers." *Arithmetic Teacher* 31 (February 1984): 3.

________. "The Rational Number Construct—Its Elements and Mechanisms." In *Recent Research on Number Learning,* edited by Thomas E. Kieren, pp. 125–49. Columbus, Ohio: ERIC Clearinghouse for Science, Mathematics, and Environmental Science, 1980.

Kouba, Vicky L., Catherine A. Brown, Thomas P. Carpenter, Mary M. Lindquist, Edward A. Silver, and Jane O. Swafford. "Results of the Fourth NAEP Assessment of Mathematics: Number, Operations, and Word Problems." *Arithmetic Teacher* 35 (April 1988): 14–19.

Lay, L. Clark. "Mental Images and Arithmetical Symbols." *Visible Language* 16 (1982): 259–74.

National Council of Teachers of Mathematics, Commission on Standards for School Mathematics. *Curriculum and Evaluation Standards for School Mathematics.* Reston, Va.: The Council, 1989.

Skemp, Richard. "Communicating Mathematics: Surface Structures and Deep Structures." *Visible Language* 16 (1982): 281–88.

9

Effective Mathematics Teaching: One Perspective

Timothy D. Kanold

MUCH of the recent research on effective mathematics teaching focuses on instruction that promotes student activity. This instructional style requires teachers to move away from lecturing and move toward monitoring their students' readiness and checking for understanding. The instructional challenge for each of us as classroom teachers is clear: Go beyond traditional teaching methods. Teaching that articulates goals, promotes strategies for solving problems, and provides student-guided practice is *more important* than the modeling of content skills.

Research on information processing notes the necessity for providing classroom activities that allow students to transfer information from short-term memory to long-term memory as they automate skills (Tobias 1982). Teachers can help with this process by reviewing, rehearsing, and elaborating on the material. Providing time for student reflection and summary in class will enhance student success. The key to the teacher's transition from using "student passivity" to "student activity" instructional methods lies in daily planning. Planning must include preparing for classroom activities that extend beyond lectures only. Seven steps for effective daily planning are given in figure 9.1.

These focal points of daily planning are postulated on two critical beliefs supported in part by research: (1) Students need to practice and process learning *actively*. They need to be actively involved in the review process, the statement of goals, the small-step instruction, and the guided practice. (2) A most critical factor for having an *immediate impact* on student achievement and performance is *monitoring* (Brandt 1984). Walking around during the classroom period is *the* crucial element. Teachers who stay glued to their desks, blackboards, or overheads promote student passivity and management problems. They also limit their ability to diagnose effectively their students' readiness and understanding. Circulating among students helps to reduce disciplinary problems, since the efforts to check for understanding

Planning for Effective Instruction

1. Begin each class with a short overview, review, or "Problem of the Day," as appropriate.
2. Begin each lesson with a short statement of goals and rationale, verbally and in writing.
3. Present new material in small stages, with time for student practice and exploration after each stage.
4. Give clear and detailed instructions.
5. Check for student understanding by posing numerous questions and giving students a lot of active, guided practice.
6. Give specific instructions for seatwork.
7. Allow continued practice in small groups until students are independent and confident.

Fig. 9.1

also monitor the behavior of all students. Thus classroom management becomes secondary. Frequent monitoring during class of students' progress serves as an important diagnostic function, provides readiness checks for the teacher, and allows the teacher to give direct and immediate feedback to students—all of which are known to affect students' achievement and attitudes (Hunter and Russel 1981).

PLANNING THE START OF CLASS

The classroom period should begin with placing on the overhead or the chalkboard one or more review questions or a problem foreshadowing new material. As the students work through this initial activity, the teacher can take attendance, check for passes, monitor students' homework, and answer individual questions in a brisk and timely manner. It is important during this time that the teacher walk quickly around the room checking all students on a regular basis rather than be drawn in by one student for a lengthy period of time. It is expected that all students will begin work on the boardwork activity on entering class, thus bringing them immediately on task. Students are expected to open their notebooks, copy down the exercise(s) or problem of the day, and begin. As the students have completed this work, the teacher then begins class with a short statement of goals, both verbally and in writing, to create an orientation to the lesson and to create the mind-set and rationale necessary for the students to be actively involved in the mathematics. As the goals are stated, list intended outcomes of the lesson. Train students to record these outcomes in their notes each day. It is important to state the outcome as an activity. To ask the students to "understand" the Pythagorean theorem is too vague. To ask them to be able to *state* the theorem, *list* two applications of the theorem, *contrast* its use with 30-60-90 triangles, or *explain* the derivation of the theorem suggests an activity. Finally, do *not* start class by "going over" homework. Reserve this activity for later in the period.

PRESENTING THE CONTENT AND CHECKING FOR UNDERSTANDING

The move into the presentation of content invites the question, "How do I check for student understanding during the classroom hour?" Most teachers are very good at modeling examples but tend to check for understanding by relying on such verbal cues as, "Did I go too fast for you?" "Isn't this an easy one?" "Okay?" "Everyone see that?" "Who doesn't understand that?" Often included is the rhetorical "Any questions?" These particular cues set up two counterproductive conditions in the classroom because (1) the teacher makes the false assumption that no response indicates that it is okay to continue, and (2) the students develop a sense of lowered self-esteem if they do respond to them, since it is an admission in front of their peers that something is not okay (Johnson 1982). Thus, teachers need to develop successful techniques and methods for checking on students' understanding throughout the classroom period. (These checks also give students an opportunity for momentary reflections on the content being learned.) There are three effective ways to check for understanding:

1. *Reflective Summaries during the Presentation of Content*

On a complex word problem, the teacher could help the students set up the initial investigation of the problem but then allow them time to discuss strategies for solving it. This is a good time to circulate among students to find out if they understand the set-up of the problem and possess the skills necessary to solve it. While walking around, the teacher is receiving and giving feedback and checking for understanding to determine the pace of the lesson. Students can also summarize major points during the class period. After completing several examples for solving systems of two equations in two variables using the elimination method, students could be asked to summarize the major steps for the solution algorithm either in writing or by discussing the process with a partner. Thus, rather than be required to memorize an algorithm that is based on isolated examples, students are given time to understand the underlying steps of the algorithm.

2. *Effective Questioning*

Effective teachers use strategies that encourage *all* students to consider the questions they are asking. Questioning opportunities exist during review, the presentation of content, guided practice, homework activities, and generally during any part of the class period in which the teacher is trying to assess group awareness, readiness, and understanding. An effective questioning cycle must allow students to listen actively both to the question as well as *to other students' responses*. A questioning cycle likely to result in this active engagement includes the following four steps:

1. Pose the question.
2. Provide "wait time" after each question to prevent student callouts.
3. Select students randomly, making certain to include all students. Call on volunteers as well as nonvolunteers.
4. Redirect the student response to other students for their judgment of correctness or for an extension of an answer.

During questioning periods, some students are reluctant to wait to be called on and like to call out an answer or response. Student callouts can be disruptive because these students then begin to control the pace and the direction of the classroom focus. This particular cycle of questioning reduces student callouts. The use of wait time allows students to think of a response before being called on (Barell 1985). A technique that also encourages wait time is to immediately follow a question with a phrase such as, "Raise your hand when you're reasonably sure of the answer." *Reasonably sure* indicates that it is okay to take a risk and possibly be wrong. The questioning sequence provided below promotes the notion of an effective question cycle.

T: Class, what would be an example of a triangle with area 12 cm^2? Please draw and label a diagram on your papers and raise your hand to respond. [Teacher monitors students as hands are raised.]

T: Only Jessica and Adam have such a triangle? Are there more? Three hands, four hands, anyone else? [Total wait time is eight to ten seconds.]

T: Okay, Roy, I noticed you chose a right triangle. Please explain your diagram to the class. [Roy did not have his hand up.]

R: I drew a right triangle with legs of length 8 cm and 3 cm.

T: How many agree with Roy's example? Raise your hands! Who disagrees with Roy's example? [This forces other students to pay attention to Roy's response.]

T: Who can prove or disprove Roy's assertion . . . Connie? [Connie had her hand up.]

C: In a right triangle, the legs represent a base and height. Thus $A = 1/2bh$ or $A = 1/2(8)(3)$, which is 12 cm^2?

T: Very good! Thank you! As I walked around I noticed all of you used a right triangle. Can you think of an example that is *not* a right triangle? [Dialogue continues.]

This questioning cycle allows teachers to know which students are "with them" during the classroom period. It also increases wait time, limits students' callouts, and promotes group feedback to a student response and positive group success on questions asked in class. This style of questioning and group checking for understanding should be used constantly during various classroom activities.

3. *Controlled Guided Practice and Monitoring*

Teachers must plan and provide multiple guided-practice opportunities for students throughout the classroom period. A guided-practice opportunity should follow the teaching of a concept or a procedure. The teacher asks all the students to attempt a particular task while the teacher monitors their work. This time can also be used to test problem-solving strategies. A teacher can give students opportunities to use higher-order thinking skills by asking them to list two or three strategies they might use in solving a problem. As the students begin to make their lists, the teacher observes the ideas they are writing down. In addition to summarizing key ideas or major points in their papers, students can also work in small groups. Guided practice is an excellent time for students to be actively involved in mathematics with each other. Students can work in pairs, with one person explaining a strategy or solution to the other. These math-communication activities can be oriented toward problem solving in activities that require students to "formulate a response," "investigate patterns in the diagram," "develop and apply a strategy for attack," or "verify and interpret results."

PLANNING FOR THE END OF CLASS

Generally one of two conditions exists at the end of a class period: either little time is left and the teacher is scrambling to give the assignment, or ten to fifteen minutes of class time remain. Many teachers feel their "job" is done when the material has been presented. However, it is important for teachers not only to give the assignment but to discuss how to do the assignment. Teachers should review instructions, make sure the directions are clear and precise, and hold the students accountable for their work (McGreal and Collins 1983). The end-of-period time can be an excellent opportunity for teachers to monitor the start of the homework and diagnose any potential problems. It is also a time *to plan* for working with those students who generally do not risk getting involved.

If time is limited, use an end-of-the-hour wrap-up activity. For example, have the students write down in their notes the three major ideas they have learned for the day. They can also talk about the strategies they have used to try to reach the major objectives of the day. *What* is to be learned and *why* they should learn it should not be kept a secret! These wrap-up activities will help students automate content as they develop their own process for learning. Also at this time homework from the previous day can be discussed, graded, corrected, and examined. Key problems can be discussed and reviewed using a group format. Accountability for student homework can be accomplished through (1) collecting notebooks, (2) brief homework quizzes, and (3) collecting isolated problems.

SUMMARY

This article suggests that each teacher use a daily lesson plan that incorporates the following processes:

1. Provide an initial activity that serves either as a review of previous lessons or as an advance organizer for the topic of the present lesson.
2. Explain to students *what* is to be learned and give a rationale for *why* it is to be learned.
3. Allow students to practice new skills and to reflect on the strategies used to perform those skills during the class period. Monitor them and provide feedback as needed.
4. Assign on a daily basis homework that promotes high success rates on using skills and rewards creative thinking on problem-solving activities.
5. Use a questioning cycle that forces students to wait before responding to a question and at the same time uses qualifying statements that allow all students to get involved actively in the question.
6. Interact with the whole class during class time and move students through discussions at a brisk pace with a high level of enthusiasm.
7. Force students to interact and communicate during part of the period.

These ideas do work. They are practical and will have a positive effect on student achievement if combined with a teaching personality imbued with an endless sense of humor and a love of mathematics!

REFERENCES

Barell, John. "You Ask the Wrong Questions!" *Educational Leadership* 42 (May 1985): 18–23.

Brandt, Ron, ed. *Effective Teaching for Higher Achievement.* Alexandria, Va.: Association for Supervision and Curriculum Development, 1984.

Hunter, Madeline, and Doug Russel. "Planning for Effective Instruction: Lesson Design." In *Increasing Your Teaching Effectiveness,* pp. 63–68. Palo Alto, Calif.: Pitman Learning, 1981.

Johnson, David R. *Every Minute Counts.* Palo Alto, Calif.: Dale Seymour Publications, 1982.

McGreal, Thomas, and Craig Collins. "Seatwork: A Perspective for Teacher Trainers and Supervisors." Unpublished mimeograph, University of Illinois, Champaign, Ill., 1983.

Tobias, Sigmund. "When Do Instructional Methods Make a Difference?" *Educational Researcher* 11 (April 1982): 4–9.

10

Beyond Problem Solving: Problem Posing

Barbara Moses
Elizabeth Bjork
E. Paul Goldenberg

"How many coins does it take to make 45¢, using just nickels?"
Nine . . .
. . . And then we are finished.

WHAT IS PROBLEM POSING?

As the problem above stands, it is fully specified: a simple question with a unique answer. But behind each such problem, no matter how limited it may seem, lurks a world of other potentially interesting problems. One strategy for getting to these problems is to ask what *kind* of information the problem gives us, what *kind* of information is unknown (and wanted), and what *kinds* of restrictions are placed on the answer. An analysis of the coin problem is given in the chart shown in figure 10.1.

	Kind of Knowledge	*Details*
Known	The sum of money	45¢
Unknown	The number of coins	
Restriction 1	A *particular* coin must be used at least once.	5¢
Restriction 2	All coins must be of the same type.	

Fig. 10.1

Beyond merely citing the seminal writing of Stephen I. Brown and Marion I. Walter, the authors wish to acknowledge the personal influence these two teachers have had on our thinking and work. This paper was prepared under partial support by the National Science Foundation, Grant Number MDR-8651637. Opinions expressed herein are those of the authors and do not represent NSF policy.

Many new problems can be generated from our original problem by changing the knowns or unknowns or the restrictions.

- What if we reversed the known and the unknown? That is, what if the *sum* were unknown and the *number* known? A new problem is born: What sum of money can be made using exactly nine nickels? And the problem is still fully specified.
- What if restriction 1 were dropped and the problem did not specify that nickels be the coin used? A new problem is born: How many coins does it take to make 45¢, using just one type of coin? The answer is not unique; it is either 9 or 45, depending on one's choice of coin. If we also reverse known and unknown in the original problem, we have yet another problem: What sum(s) of money can be made using exactly nine coins, all of the same type?
- What if restriction 2 were dropped? A new problem is born: How many coins does it take to make 45¢ using at least one nickel? Again, the answer is not unique; there are seven different ways of making 45¢ using exactly one nickel, and lots more ways using more nickels.
- The implied domain is *all* standard American coins: 1¢, 5¢, 10¢, 25¢, and so on. What if we changed domains and used a different set of coins?

  *What if we didn't have pennies, but otherwise used American coinage? What sums of money could we no longer construct?

  *What if we didn't have pennies, but did have a 3¢ coin (and other American coinage)? What sums of money could we no longer construct? (We would no longer have exact change for 1¢, 2¢, 4¢, and 7¢.) Even though we could not buy certain objects with exact change, perhaps some of them could be bought if we paid more and collected the right change (e.g., paid 10¢ and received 3¢ change). Which ones? (All of them)

HOW DO WE POSE PROBLEMS?

Identifying and Changing Constraints

> **Principle 1:** Have students learn to focus their attention on *known, unknown,* and *restrictions.* Then consider the following question: What if different things were known and unknown? What if the restrictions were changed?

The most mundane of problems can become richer if instead of just asking, "How do I solve it?" we first ask, "What's this problem all about?"

Principle 1 organizes a child's exploration of what a problem is about and suggests a way to create a new problem.

Brown and Walter (1970, 1983) maintain that this approach is not only learnable but teachable. Students can be taught to expect that mathematical statements and problems, like the one that we explored at the beginning of this paper, include something *known,* something *unknown,* and (often) some *restrictions*. Sometimes these restrictions are subtle: by thinking about stamps instead of coins, we might notice that there are more denominations than 1¢, 5¢, 10¢, and so on. Removing a restriction shows how a problem is part of a larger class of related problems rather than an isolated exercise. Initially, the teacher bears most of the responsibility for helping students see the classes from which each problem instance is drawn. In time, students do more and more on their own, building a repertoire of associations for creative problem posing and effective problem *solving*.

Looking at Familiar Things in Strange Ways

Principle 2: Begin in comfortable mathematical territory.

By starting in a context that is sufficiently familiar, even very young children—with some encouragement and modeling by the teacher—can list attributes and change the constraints of a problem. This kind of thinking should be begun in the earliest grades to encourage a problem-posing attitude.

For young children, manipulatives often help to make a mathematical territory feel familiar. In our first example, we used coins to give a homey feel to what might otherwise have been a very abstract and arbitrary set of restrictions about addition problems. Brown and Walter (1983) show how even a common manipulative like a geoboard can be used for attribute finding. As they are, geoboards have square borders, pegs at lattice points on a square lattice, and finite size. What if one or more of these characteristics were changed? And what new questions does a change in characteristics lead one to ask?

For example, what if the geoboard's outline were an oblong or a circle instead of a square? We might then ask what such a change of shape does to the number of pegs on the board. For example, how many fewer pegs are on a diameter-2 *circular* segment of a geoboard than on the entire 2×2 *square* geoboard? Figure 10.2 shows circular segments of 2×2 through 5×5 geoboards. On the smallest of these geoboards, 4 pegs lie outside the circle. The others exclude 12, 12, and 20 pegs, respectively.

Varying the attributes and asking new questions leads to a host of discoveries.

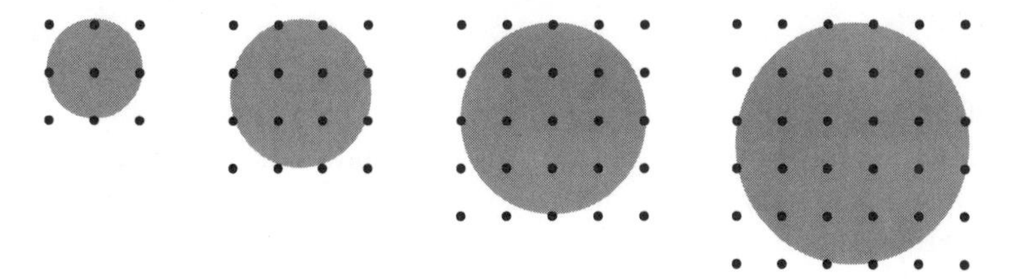

Fig. 10.2

Using Ambiguity: Necessary but Insufficient Conditions

> **Principle 3:** Encourage students to use ambiguity to create new questions and problems.

Sometimes children find it easier to express what they *want to know* about a mystery object or an ambiguous statement than to state what they *do know* about something that is completely revealed to them. Hold up a box with some coins in it, tell the students it contains coins, jingle the coins, and encourage students to ask questions about what's in the box. Record all questions. If necessary, stimulate new trains of thought by suggesting questions like "Are they all U.S. coins?"

A common teaching technique aimed at clear communication uses this idea in reverse. The teacher asks students for a definition of, say, a triangle and deliberately misinterprets an incomplete student response such as "It has three lines" by producing a drawing like that shown in figure 10.3. This is problem posing par excellence—keeping one feature (the student's specification) fixed while relaxing the other constraints—and so it is a perfect opportunity to model that problem-posing technique. Each incomplete student specification, rather than being treated as wrong, is seen as one essential feature—perhaps insufficient but absolutely necessary. As students get the idea, the teacher may do less of the problem posing and turn that task back to the students: How can we draw a figure that conforms with the feature just mentioned but is not a triangle? Or in the language of "what if," What if it had three lines but was not a triangle?

Fig. 10.3

The goal should be not just to "catch" the ambiguity as if it is merely an error but to *use* it productively. This approach not only de-emphasizes the inadequacy of one student's incomplete specification but has the added advantage of focusing all students' attention directly on the unspecified attributes, helping them learn to recognize those attributes. The amusement value of the perverse drawings and the puzzle are maintained, as well as the goal of helping students use more precise communication.

Ambiguity leaves room for curiosity, imagination, and generating one's own ideas. Ambiguity is also inevitable. Where it arises, we should make it useful rather than a failure.

Making the Domain Explicit

Principle 4: Teach the idea of domain from the earliest grades, encouraging children to "play the same (mathematical) game with a different set of pieces."

Games may help students look at constraints creatively. In some kinds of games (especially ones like chess, nim, go, and 21, which we see as very mathematical), there are playing pieces (pawns, counters, cards, etc.), rules by which these pieces may be manipulated (moving, taking, counting, etc.), and goals to achieve (a particular configuration of pieces, a specified sum, the most or least, being last to move, etc.). By *domain* we mean the mathematical "objects"—they may be specific numbers, geometric shapes, functions, or other mathematical abstractions—that we decide to include in our game.

There are always opportunities to explore different domains that remain within the child's mathematical competence, and from the earliest grades students should be taught to "try the same game with a different set of pieces." Any mathematical task that a child can perform in one domain (e.g., counting numbers) can be explored within a subset of that domain (e.g., evens, or the set {1, 4, 7, 10, 13, . . .}). Often, the question "What numbers am I allowed to use?" is sufficient to call the idea of domain to mind. For example, in the problem *Name two numbers whose product is 12,* do we allow all numbers, only integers, only evens, only odds? (In the case of "only odds," of course, the problem has no answers.)

HOW CAN WE FOSTER CREATIVE PROBLEM POSING?

The teacher is the essential ingredient. It is the teacher who sets the context by helping students learn how to open up one problem to reveal the others it suggests. And it is also the teacher who establishes a classroom climate conducive to spontaneous and productive inquiry in several ways: modeling the process personally by wondering openly *with* the students, fostering the free exchange of ideas and actively encouraging collaboration among students, honoring students' spontaneous what-ifs and conjectures, and being as interested in *how* students thought about a problem as in *what* they came up with.

The following two strategies can foster problem posing among students.

Use Problems in the Textbook as a Basis for Problem Posing

Using techniques such as those outlined above, the teacher can regularly pick a problem that can be enriched through problem posing. One might begin with the best problems in the text, select problems from supplementary sources (e.g., Greenes et al. [1980], Lane County [1981], and Ohio Department of Education [1982]), or deliberately pick a not-so-interesting problem to explore and subsequently think out loud with the students.

Avoid Questions That Have Unique Answers

Problems that have more than one solution tend to foster a problem-posing mind-set because they are not as limited as one-answer problems. For example, the questions

"How many coins does it take to make 45¢, using *at least one* nickel?"

or, more broadly,

"How many ways can you make exactly 45¢ with U.S. coins?"

provide a richer context for problem posing than the following question:

"How many coins does it take to make 45¢, using *just* nickels?"

Classroom Climate

Although most students are not used to being problem posers, they can learn to be if their teachers model the process for them and create an environment in which they feel free to pose their own problems. We offer the following suggestions.

- *Let students choose what problems they try to solve.* Good problem posers will raise more problems than time permits them to solve. Also, some problems will generate more interest and curiosity than others, and so individual taste and choice will play a part in who follows up which problem and to what extent. Students may also raise problems that they *cannot* solve. Students feel free to pose problems when they do not fear being embarrassed if they invent one that is too hard for them. Some of the most important discoveries in the history of mathematics were inspired by problems that still remain unsolved.

- *Place no time pressure on solving a problem.* Good problems take time to explore and yield many new interesting ideas. For both reasons, good problem posing requires lots of exploration time. Student problems that remain unsolved for whatever reason can resurface from time to time, stimulating curiosity and further thought, and should be seen as challenges, not failures.

• *Brainstorm with the students; encourage communication and collaboration.* Students feel most free to risk posing new problems when they and the teacher work *together* as collaborating partners in problem posing. Encourage the productive exchange of ideas among groups of children and remain flexible about the direction of the lesson and the stream of student inquiry and initiation. Mathematical content remains essential, but when students and teacher focus on the process of solving problems and the challenge of posing new ones, the emphasis shifts from *acquiring* facts to *using* them.

Collaboration among students has many benefits. For one thing, it provides a context in which students must develop ways of communicating about mathematics and build a vocabulary for their mathematical ideas. The *Curriculum and Evaluation Standards for School Mathematics* (NCTM 1989, p. 6) emphasizes that "as students communicate their ideas, they learn to clarify, refine, and consolidate their thinking." Collaboration and communication are particularly important for problem posing, for as students share their different ideas, questions, and perspectives, they fertilize the development of new ideas, questions, and perspectives.

Record and honor spontaneous what-ifs and conjectures. Naming an idea after its inventor—Rachel's method, the Jonathan conjecture—can be a great way of honoring the inventor and encouraging others.

Using Technology to Promote Problem Posing

Computers and calculators free students and teachers alike from tedious computational tasks, from repetitive numerical and geometric manipulations, and from overwhelming memory jobs. Because the computer calculates and recalculates so quickly, students can easily ask and explore the what-if questions with relatively little tedious computation. In this way, appropriate use of technology can foster and enhance problem posing by students.

In one fourth-grade classroom, the students were asked to find an average score for their recent mathematics test by using a piece of graphing software. Students were able to enter the test data quickly, display bar graphs, and compute the mean score of the class. When a new child joined the class, the students were able to add the new student's score and see how it affected the bar graph and the mean score for the class. They could also ask many what-if questions; for example: "What if two students dropped out of the class?" "What if the teacher had made a five-point error in each student's score?" "How many scores need to be adjusted to have an average of 90?"

The computer allows students and teachers to explore domains that were previously unwieldy or unavailable in a pencil-and-paper world. Data can easily be collected, and the computer will display them in many forms, allowing the student and teacher to make predictions and ask new questions about the effects of changing any one of the conditions. New problems are easily generated because the tool displays graphically that which the students are exploring.

Skilled communication contributes significantly to mathematical learning. Students need to be encouraged to speak and write mathematics and learn how to sense when words are not as effective as pictures, diagrams, symbols, and graphs. The computer seems naturally to encourage collaborative experimentation and the sharing of results of problem posing and solving. When it links graphical and symbolic representations, it also helps students see when each representation is appropriate and then learn to choose among them.

CLASSROOM EXAMPLE: A LATTICE SETTING

Because of our strong belief that problem posing is central to the learning of mathematics, we have begun writing K–6 curricular materials in which problem posing pervades all the activities. This section describes one area of the *Reckoning with Mathematics* curriculum and the way problem posing arose by (1) using interesting problems, (2) having the teacher model good problem-posing behavior, and (3) letting the students follow their own interests. Notice how the technology enhanced the problem-posing process.

We showed a class of fourth graders a ten-lattice (Page 1965, Goldenberg 1970) as a way of looking at number patterns. Shown an incomplete matrix of ten columns (fig. 10.4), students were asked to fill in the missing numbers. What number should be next to 32? What number should be above 32? Below 99?

90	91	92							99
•									
•									
•									
•									
•									
30	31	32							39
20	21	22	23	24	25	26	27	28	29
10	11	12	13	14	15	16	17	18	19
0	1	2	3	4	5	6	7	8	9

Fig. 10.4

The teacher explained that although the lattice drawn on the board ended at 99, the lattice actually continued. For how long? "Until you die!" asserted one student. The chalkboard is a static medium, but using a piece of prototype software designed by Education Development Center for this module, the teacher and students were able to generate any lattice and scroll it up and down to show how it continued "forever" in both directions.

Because of the rich environment of the lattice and the familiarity of the domain, the teacher's role at this point was to pose a few open-ended questions. "What patterns do you see? How many rows up would you have to go to reach numbers in the thousands? What could we put below zero?" The question "How would you describe the lattice to someone who had never seen one?" is an attribute-finding question from which the children could begin to pose problems. At the teacher's suggestion that other lattices could be made, the entire class got into the act. "What if we started at 2? Could we start at -1? Could we make the numbers go down [decrease]?"

With the software that built lattices to the students' specifications, the children invented lattices as quickly as they could generate the ideas. They posed new problems:

- What is the difference between a lattice that starts at 3 and one that starts at 2?
- What happens if the lattice has only nine columns? Five columns? One column?
- What would the lattice look like that has ten columns and only even numbers?
- Could we create a ten-column lattice with the number 150 in the third row?
- Is it possible to make a lattice where halves are included?
- How many rows up would I need to go before reaching the number 542 in the ten-lattice? (This was easy to determine because the software permitted the children to move up many rows at a time.) How does this compare to the numbers of rows needed before reaching the number 542 in a twenty-column lattice?

Although the wording is ours, these were *student*-generated questions. Each new question inspired two or three new ones, and students were inventing and doing mathematics as they speculated on the answers. The freedom of the classroom led to a deeper understanding of the nature of number.

WHY IS PROBLEM POSING GOOD FOR US?

An orientation toward *posing* new problems can be said to be the very heart of learning mathematics. Learning is a creative act: we learn not by absorbing but by *constructing* our knowledge. And we learn mathematics particularly well when we are actively engaged in creating not only the solution strategies but the problems that demand them.

What else is problem posing good for? It probably lessens mathematical anxiety. In certain classes where problem solving consisted of finding a solution to a teacher-posed problem, we noticed a great deal of anxiety among the students. There was a fear of being wrong or of thinking up foolish ideas. However, in a problem-posing environment, there is no one right answer. Students were willing to take risks, to pose what they considered to be interesting variations of the problem. As Brown and Walter (1983) point out, mathematics became less "intimidating."

Problem posing also helps us to notice our misconceptions and preconceptions (see example in principle 3) and may, in that way, help us become better consumers in today's world. Presented thoughtfully, the problem-posing spirit can augment the rest of our teaching about the difference between the literal claims in advertising or political rhetoric ("it has three lines") and what we thought the product or promise really was (a triangle). By considering creative misinterpretations in the context of problem posing, we develop this new awareness.

Finally, problem posing helps to foster group learning rather than competition against the other members of the class.

In a world where the necessary mathematical skills are changing so rapidly, it will be the good problem poser who will be the best prepared.

REFERENCES

Brown, Stephen I., and Marion I. Walter. *The Art of Problem Posing*. Hillsdale, N.J.: Lawrence Erlbaum Associates, 1983.

________. "What-If-Not? An Elaboration and Second Illustration." *Mathematics Teaching* 51 (1970): 9–17.

Goldenberg, E. Paul. "Scrutinizing Number Charts." *Arithmetic Teacher* 17 (December 1970): 645–53.

Greenes, Carole, George Immerzeel, Earl Ockenga, Linda Schulman, and Rika Spungin. *Techniques of Problem Solving*. Palo Alto, Calif.: Dale Seymour Publications, 1980.

Lane County Mathematics Project. *Problem Solving in Mathematics*. New Rochelle, N.Y.: Cuisenaire Company of America, 1981.

National Council of Teachers of Mathematics. *Curriculum and Evaluation Standards for School Mathematics*. Reston, Va.: The Council, 1989.

Ohio Department of Education. *Problem Solving: A Basic Mathematics Goal*. Palo Alto, Calif.: Dale Seymour Publications, 1982.

Page, David. *Maneuvers on Lattices*. Newton, Mass.: Education Development Center, 1965.

11

Writing as a Tool for Teaching Mathematics: The Silent Revolution

Aggie Azzolino

BECAUSE writing is a way of clarifying and refining one's own thoughts as well as communicating with others, mathematics has a rich history of using writing to learn. Mathematicians write both for themselves, making concrete that which was only thought, and to colleagues and friends, communicating through letters, papers, and journal articles. Teachers write for themselves and for their students in textbooks, lecture notes, lesson plans, and tests. In teaching a new course, we must often learn new material on our own: writing is a way to do this. Students of mathematics can also write for themselves and for their teachers.

As you read through the following instructional purposes of writing, check goals you have for yourself or your students. Writing can—

- demand participation of the student;
- help the student to summarize, organize, relate, and associate ideas;
- provide an opportunity for a student to define, discuss, or describe an idea or concept;
- permit the student to experiment with, create, or discover mathematics independently;
- encourage the personalization, assimilation, and accommodation of the mathematics being taught;
- assist the student in reviewing, refocusing, and reconsidering topics either recently studied or considered long ago;
- assist in recording and retaining mathematical procedures, algorithms, and concepts for future use;
- assist in the translating or decoding of mathematical notation;
- assist in symbolizing or coding with proper notation;
- help the teacher diagnose a student's misconceptions and problems;

- provide an appropriate vehicle for the student to express and focus on negative feelings and frustrations as well as to emote and rejoice in the beauty of mathematics;
- assist in the reading, summarizing, or evaluating of texts;
- improve your teaching (Azzolino 1987);
- collect evidence for research.

GETTING STARTED IN USING WRITING AS A TEACHING TOOL

How might these goals be achieved in your mathematics classroom? *Don't glance ahead.* Take pencil in hand and write the ways in which you might use writing to achieve the goals. Compare your list with figure 11.1, appending any items you've forgotten; then use the revised list as a core of the writing assignments you consider best for your teaching style. Consider the following suggestions as you refine and expand your core.

Using writing as a teaching tool is a very personal thing. There are no universals. If an assignment doesn't work for you, then change it, rework it, save it for another time, or don't use it.

Do not dwell on spelling, punctuation, or grammar. Edit the work yourself or return the work to its author for revision. Don't collect every writing assignment. Some assignments should be considered practice just as other kinds of exercises are given in class as practice. Writing is not the end but means for teaching mathematics.

IN-CLASS WRITING

Class writing activities include completion, lead sentences, warm-ups, rewording, word banks, and debriefing. These methods are discussed in the following sections.

Completion, Lead Sentences, and Warm-ups

Completion means starting a sentence and having students complete it. This technique is appropriate for summarizing, comparing, analyzing, and expressing feelings. Consider the following examples:

> *Complete:* Today we learned about something called __________. Absolute value means __________. Lines and line segments are similar because __________. Lines and line segments are different because __________. The most confusing thing today was __________. Extraneous roots exist when solving this type of equation because __________.

Ways to Use Writing in the Mathematics Curriculum

Out of Class

- Writing cheat sheets for tests
- Writing term papers
- Completing statistics projects
- Writing abstracts
- Critiquing readings
- Writing possible test questions
- Making journal entries—

  open-ended
  nongraded
  free format

- Writing technical papers
- Writing dictionaries
- Summarizing readings
- Completing take-home tests
- Writing word problems
- Writing letters to the teacher
- Completing questionbooks (Azzolino and Roth 1987):

  specific questions
  graded and regraded
  structured format

In Class

- Writing drafts of papers
- Beginning class with warm-ups
- Note taking
- Making journal entries
- Writing, exchanging, and critiquing word problems
- Doing short writings of sentences, phrases, lists, paragraphs
- Debriefings after an important point or procedure
- Debriefings at the end of the class

On Tests

- Writing proofs stated in English or symbolically
- Writing or completing definitions
- Correcting false statements in true-false questions
- Explaining procedures
- Writing essays

Fig. 11.1

A complete sentence is not required. Writing a word or phrase is valuable.

Posing the same question without requiring a written response *permits* but does not *demand* interaction or involvement. Requiring written responses requires more time, but the length of time depends on what the teacher decides to do with responses. The teacher could have students share their responses orally, exchange papers, give their papers to the teacher, or continue with the presentation.

Jot down a completion exercise you have used or could use with your class. Reread your exercise. If adding an extra sentence before the completion sentence would make the exercise easier, then use the *lead-sentence* technique.

In using a lead sentence, one makes a statement, then has students write a second sentence or a paragraph. This technique is helpful for explaining an idea, writing a definition, or creating a word problem. For example:

1. The discriminant is useful in determining the kinds of roots of a quadratic equation. The discriminant ________.
2. The factors of 18 are 1, 2, 3, 6, 9, and 18. Factors are numbers that ________.
3. The only natural number that is neither prime nor composite is 1. A prime number is ________.

Completion and lead-sentence formats are useful when asking students to focus on a small part of a larger picture.

Warm-ups expedite the consideration of the larger picture. Students are asked a question that can be answered without much thought and then asked a second question on the same topic that requires more thought. An example follows:

1. Complete: Lines that are parallel have the same ________.
2. State the equation of a line that is parallel to $y = -4x + 8$.

Sure, you could simply ask question 2, but asking question 1 first and then question 2 decreases the time needed to answer question 2. We often ask questions hoping each student will form a response, but often the same group of students volunteers verbal responses. By asking each student to respond in written form, each student gets a shot at answering the "harder" question and getting it right. Asking for written responses encourages each student to answer.

Rewording, Word Banks, and Debriefing

Rewording means having students take a statement, definition, or procedure and reword or rewrite it using different terms or using their own words. This technique is a good diagnostic, a good reinforcer, and a good way of personalizing content. We often ask students if they heard or understood what we've said, and they nod their heads yes. Rewording is a way of demonstrating understanding. Examples of short in-class, longer in-class, and out-of-class activities follow.

1. Make a statement; ask students to write exactly what you just said, quoting you as closely as possible.
2. Make a statement; ask students to restate in their own words what you just said.
3. After asking a question, ask students to restate the same question in another way.
4. Write a formula or equation; have the students copy the formula and label each variable with a word, using arrows to point to each variable.
5. Write an equation; have students translate the equation into a sentence in English.
6. Have students read a familiar question or problem; ask them to write their own question like the first question and then answer this question.
7. Have students write an easier, or more difficult, story problem. Figure 11.2 gives an example of a lesson-long format (Azzolino 1987).

Writing Your Own Word Problems

1. Write your own word problem similar to the ones you've been doing in class.
2. Change the problem you wrote in question 1 so that it is an easier problem. To make a problem easier:
   - Write fewer words.
   - Make the sentence structure simpler.
   - Make the numbers smaller.
   - Make the figures less complicated.
   - Require fewer steps or operations to solve the problem.
   - Eliminate extraneous data.
   - Present information in the order in which it will be used.
   - Introduce manipulatives (charts, diagrams, concrete objects).
3. Change the problem you wrote in question 1 so that it is a harder problem. To do this, you might try the opposite of one or more of the suggestions listed in #2.

Fig. 11.2

Word banks and debriefing are wonderful tools. A *word bank* is a list, or bank, of words. To use this technique, ask students to write a sentence or paragraph using two, three, or all the words in the bank. This technique is helpful for writing definitions, relating two ideas (Geeslin 1977), writing long explanations, or theorizing about related or apparently unrelated thoughts. Examples follow:

1. Use the words *one* and *is less than* in a true sentence.
2. Use the words *intercepted, center,* and *angle* to help you write a definition of a central angle.
3. Write a question containing one or more of these words: *right angle, perpendicular, side, hypotenuse, parallel.*
4. Write a paragraph containing as many of the following words as possible: *slope, x-intercept, coordinate, abscissa, ordinate, point, tangent, parallel, perpendicular, y-intercept.*
5. Write a true sentence using the words *rectangle* and *square* but without using the word *always.*
6. Write a paragraph about pyramids and cones.

Longer in-class assignments, whole lessons really, are possible using individual techniques and collections of the techniques above. Figure 11.3 presents material appropriate for the early elementary grades (Azzolino 1987). Notice that the exercises can also be used for short in-class activities on many different days.

Debriefing is an important tactic for reinforcing student learning and obtaining student feedback. Given the completion of a procedure, reading, discussion, or lecture, have the students list important ideas, the steps in a procedure, new words, and state the most important idea discussed. Activities include the following:

1. After you complete a procedure or algorithm, ask students to explain or state what you did in a specific step.
2. After completing a procedure or algorithm, ask students to list the steps used to complete the algorithm.
3. Have students summarize the lecture in a standard format or have them list the major topics of lecture and key words for each topic.
4. Have students list the mistakes they made on their homework, their last test, or in class.
5. Have students list four topics from a chapter in their text and write a summary of one of them.

It should be noted that this sort of summarizing can be habit forming and beneficial. It encourages students to reconsider and review what has come before. It does not take the place of taking notes, but it does help the students to organize the material presented. Figure 11.4 demonstrates one pretest debriefing format (Azzolino 1987).

Speaking of Numbers

zero	one	two	three	nine	ten	eleven	twelve
odd	even	is greater than		is less than		sum	plus

Use the words above to complete your own sentences. Make true sentences.

1. Six is less than ______________.

2. ____________________ is greater than ____________________.

3. ______________, ______________, and ______________ are even numbers.

4. The sum of seven and two is ______________.

Use these words in a true sentence.

5. sum, two ____

6. odd ____

Who Am I?

12. I'm seven more than nine. Who am I? ______________

13. I'm an even number greater than six but less than ten. Who am I? _____

14. I'm five times the size of ten. Who am I? ______________

15. I'm one-third the size of twelve. Who am I? ______________

16. I'm 2 less than 20. Who am I? ______________

Complete your own "Who am I?" question.

17. I'm ____

    Who am I? ______________

18. I'm ____

    Who am I? ______________

Fig. 11.3

Pretest Review Sheet

1. You have been given the points (3, -4) and (2, 6). List at least three things your teacher might ask you to do on the next test using these two points.

2. Using one activity that you listed in #1, write a question or problem on this task as your teacher would ask it.

3. Use the same question you wrote in #2, but change its wording so the question sounds different but means exactly the same thing.

4. List some things you must remember when answering this type of question or doing this type of problem. Don't answer the question you wrote.

Fig. 11.4

WHEN SHOULD STUDENTS WRITE?

The times and places to use writing as a teaching technique are the times that are comfortable for the teacher. As you experiment with using writing, you will find more times that are appropriate. Some teachers use writing at the beginning of class when the students are still settling down and before unpacking their texts and materials. Word banks are great for this time. When you get a room full of blank stares, ask students to reword or debrief. To punctuate a topic or when in transition between activities, try debriefing, completion, or a word-bank activity.

At the end of class, even with just a minute or two left, debriefing is really useful. Writing is a way for students to summarize ideas for themselves.

Writing on Tests

Time is the biggest obstacle to having students complete a writing task on a test. Simplifying the writing task makes writing a feasible testing technique. On tests, writing lists and summarizing topics take less time than writing definitions. Require brainstorming or quick answers rather than something that requires scrutiny or precise editing.

Do not ask students to write definitions on tests. If they have been given a good definition to memorize, then merely asking for it condones parroting. Instead, ask students to *apply* a definition or show why the definition does not hold. Since writing a new definition takes thought and time, take -home tests, out-of-class assignments, or in-class assignments—especially with word banks—are better places to ask students to write definitions.

Asking students to explain a procedure can be an excellent test exercise. Students can produce this type of writing under test conditions particularly, if debriefing activities either in or out of class have been provided. Students may need to be taught what to expect. Give them practice before asking them to do similar work in a test situation. Assign exercises that incorporate feedback from the teacher and revision by the student.

If you ask students to write an essay, consider permitting them to bring handwritten outlines, perhaps detailed ones, to the test. A good idea might be to require a rough draft of the possible essay questions to be written before requiring a final draft on a test. Consider using a word bank to stimulate inclusion of points you feel are important. Remember the things you need to do before you are ready to write about some topic in mathematics, then provide these same things for your students before you request them to write an essay.

Writing has the potential of expanding the variety of questions asked on a test. Previously, the verbs used on a test might include *complete, correct, create, define, describe, determine, establish, evaluate, expand, factor, fill in, graph, list, match, multiply, pick/choose, prove, rationalize, reduce, simplify, sketch, solve, state,* or *translate.* Once writing has been used as a testing tool, the verbs might include *analyze, compare, contrast, explain, hypothesize, justify, read and explain, relate, restate, reword, summarize, support, suppose.*

CONCLUSION

Please complete one more writing exercise. Debrief yourself by listing the new words you read in this chapter and by stating two ideas or thoughts worth remembering.

REFERENCES

Azzolino, Agnes. *How to Use Writing to Teach Mathematics.* Keyport, N.J.: Mathematical Concepts, 1987.

Azzolino, Agnes, and Robert G. Roth, "Questionbooks: Using Writing to Learn Mathematics." *AMATYC Review* 9 (Fall/Winter 1987): 41–49.

Geeslin, William E. "Using Writing about Mathematics as a Teaching Technique." *Mathematics Teacher* 70 (February 1977): 112–15.

12

Motivation: An Essential Component of Mathematics Instruction

Emma E. Holmes

MOTIVATION fuels mathematical learning. If children are motivated, they attend to instruction, strive for meaning, and persevere when difficulties arise. Competent teachers, effective instructional models, and thought-provoking activities guide the process, but children must first be motivated to learn mathematics.

Good mathematics teachers rely on both extrinsic and intrinsic motivation to promote learning. Such external incentives as grades or awards foster extrinsic motivation, whereas intrinsic motivation stems from such internal factors as interest or a desire to understand. Contemporary cognitive theories of motivation emphasize *intrinsic* motivation. Cognitive theorists believe internal factors influence a desire to learn and to invest in learning experiences more than external forces.

A FRAMEWORK FOR MOTIVATING MATHEMATICS LEARNING

Contemporary cognitive theories of motivation postulate that behavior is influenced by what we think. Thought is the internal factor governing motivation (Ames and Ames 1984). This central idea supplies a framework for stimulating children's mathematics learning: Motivation is influenced by our thoughts. In a school setting, motivating thoughts relate to tasks and ourselves as learners. This section discusses cognitive theories of motivation that deal with the theme of the framework from different perspectives and thus gives background information for interpreting it.

Achievement motivation. Individuals who are motivated by achievement *set goals and exert effort* to attain these goals. Children with *learning* goals want to comprehend; they accept the challenge of learning and persist in the face of difficulties. Those with *performance* goals think about getting the right answer and choose easy tasks as a way to demonstrate achievement. They also give up when success is not readily attained.

Children with learning goals are much like those Holt (1982) calls thinkers. Thinkers are task-involved students who are interested in understanding, check answers to mathematics problems, and persevere until answers make sense. Holt contrasts thinkers with producers. Producers are interested only in obtaining right answers, and they develop strategies that center on saying or doing what they think teachers will accept.

DeCharms (1984) believes the school can enhance the achievement motivation of learners by teaching them to set goals and take responsibility for their own learning. Children in his program for enhancing motivation demonstrated higher achievement than those in the control group in several content areas, including elementary school mathematics.

Attribution theory. Attribution theorists (c.f., Weiner 1984) examine the explanations individuals give for success and failure and relate these attributions to achievement motivation. For example, certain students may think that their success in mathematics is due to high ability or effort. These students will be motivated to learn mathematics. Other students may think that their failure to learn mathematics is due to low ability or the difficulty of the material. They will not expect to learn mathematics and will not be motivated to study the subject. These students can be encouraged to learn by helping them accept the notion that success is related to effort.

Cognitive evaluation theory. The notion that individuals seek challenge, competence, and autonomy is central to cognitive evaluation theory (Ryan, Connell, and Deci 1985; Deci and Porac 1978). According to proponents of the theory, intrinsic motivation is enhanced by events that enable students to perceive that they are competent and self-determined.

Cognitive evaluation theorists believe the type of feedback given to students affects motivation. If students think that feedback is controlling and stressing goals that are external to them, intrinsic motivation deteriorates. However, if students perceive feedback as informational and feel they can use it to develop competence, intrinsic motivation will be enhanced.

Cohen (1985) notes that the manner in which a teacher gives feedback influences how it is interpreted by learners. If the feedback focuses attention on how the teacher feels and is thus controlling ("I am proud of. . . .") rather than stressing meaning to the learner ("you must feel proud of. . . ."), the feedback will not enable the learner to feel autonomous and responsible, and it will not enhance intrinsic motivation.

This section has reviewed ideas from cognitive theories of motivation that form the basis for a framework for enhancing mathematics learning in the elementary school. The next section describes the guidelines and practices derived from this framework.

MOTIVATING CHILDREN TO LEARN MATHEMATICS

This section discusses four guidelines for incorporating theories of motivation in elementary school mathematics instruction.

Helping Students Experience Success and Competency

Success in mathematics demonstrates competency. Thinking that one is competent enhances intrinsic motivation. Following are two practical approaches to enable learners to experience success together with ideas for helping children realize they are competent.

1. Help children generate knowledge. Wittrock (1979) points out that generating knowledge improves learning and influences motivation. Generating knowledge involves making inferences and applying ideas. Two specific activities showing children involved in generating ideas follow:

- Young children measure the length of the sides of a number of cardboard triangles and then decide they can group the triangles by the number of sides of equal length.

- Older students discover that the sum of the measures of the angles of an n-sided polygon is $180° \times (n - 2)$ by generalizing from the pattern established by determining the sum of the measures of the angles of 3-, 4-, 5-, and 6-sided polygons.

When students make keen observations, draw valid conclusions, and make reasonable inferences, they should be told they are competent in mathematics. Children may become aware of their competency only if teachers interpret for them the significance of what they have accomplished.

2. Teach strategies for comprehending ideas and solving problems. If students know learning strategies, they can approach learning with expectations of success. One such strategy is *visualizing* (Resnick and Ford 1981). Visualizing involves using mental images in thinking. Children learn visualizing from work in the enactive and iconic modes of representing experience. Using concrete materials and pictorial representations enables students to build a store of images to use in subsequent work. Work with physical and pictorial models of mathematical ideas, therefore, helps learners develop visualizing strategies.

Children are helped to learn visualizing as a strategy for solving problems by acting out or making sketches of problem situations. Such activities should be followed by encouraging children to make a "mind picture" of each problem situation enacted or represented pictorially.

Other strategies useful for understanding and solving problems include *guess and test, work backwards, reason,* and *use organized lists.* These strategies are often taught through modeling by the teacher.

After students learn to match strategies to problems, they can be asked to explain how they knew what strategies to use. The teacher should point out insightful comments and help children realize that they have become competent in problem solving.

Helping Students Internalize Learning Goals

Goals can be motivating and can help focus students' thoughts on achieving. Learners with learning goals want to make sense of mathematics, develop skill, and solve problems. They persevere to attain these goals. They concentrate on the task to be done. Initial failure does not defeat them; they keep trying.

To help pupils internalize learning goals, teachers should help them find satisfaction in learning, stress the value of doing mathematics, and make lessons meaningful and challenging. The following discussion gives three suggestions, with examples, to help children focus on learning goals.

1. Use cooperative learning arrangements. Cooperative learning provides opportunities for groups to study subject matter or complete assignments (see the article by Neil Davidson in this book). Children generally enjoy working together and encourage each other to complete a task.

A variety of activities can be planned for cooperative learning. Group problem solving helps learners develop problem-solving skills. Students can cooperate on projects such as preparing a bulletin board to show the hierarchical classification of geometric figures. They can work cooperatively to plan and carry out interviewing, record and tabulate data, graph the data, and make inferences and raise questions about their graph, such as, "What can we predict from this graph?"

2. Emphasize the value of mathematics. Seeing the value of knowledge promotes learning goals. Children learn to value mathematics as they discover that it is a necessary tool for dealing with both life and school situations, and they develop high regard for mathematics as they become aware of its role in the development of society. The use of mathematics in solving practical problems and its relation to social science and science are themes explored at all age levels.

If students value mathematics, they are also disposed toward it and think and act in a manner that demonstrates they believe in the reasonableness of mathematics (National Council of Teachers of Mathematics 1989). Therefore, activities that help learners experience mathematics as a reasonable system of thought help them value mathematics. Such activities can stress alternative approaches to completing assignments or emphasize the thoughtful exploration of ideas. Three examples of activities to promote a favorable disposition toward mathematics follow:

- Students explore different arrangements of six square tiles that share common sides to make figures each with a perimeter of 14 units.
- Students generate alternative ways to find products of multidigit factors (e.g., $28 \times 45 = 4 \times 7 \times 5 \times 9$).
- Students keep journals of their approaches to problem solving as discussed in Azzolino's article in this volume.

3. *Ask open questions.* Teachers should ask open questions to enable students to explain and reflect on their thinking (Holmes 1985). These types of questions often begin with "Why did you . . . ?" or "How did you . . . ?" An example is "What did you think when you found the area of this shape made on the geoboard?"

Related to open questions are the problem-posing questions described by Brown and Walters (1983) and illustrated in the article by Moses, Bjork, and Goldenberg in this yearbook. Such questions are asked by learners to explore a given or to challenge a given. After generating problem-posing questions, learners become involved in analyzing and expanding their questions; in so doing, new insights occur.

Helping Learners Experience Autonomy and Be Responsible

Feelings of autonomy can result from independent reflection on, and analysis of, a problem until a solution is found. Making choices also enables students to feel self-determined and responsible. However, to insure that sound choices are made, teachers usually ask children to select activities from a group designated by the teacher. For example, when the goal is to study basic facts to insure immediate recall, students can choose to study with a partner, write or review facts independently, or practice facts using a computer program.

Opportunities for children to be autonomous should be accompanied by discussions on responsibility. Students need to share their ideas about the meaning of responsibility. In these discussions the teacher will probably need to bring up the notion that being responsible includes being an originator, that is, initiating behaviors that characterize responsibility. Discussions should also include giving examples of responsible behavior. These examples can be posted on charts to use in evaluating the growth of responsible behavior.

Teachers can show they believe children can be responsible by having the students check their names off when homework is handed in on time, or by asking them to evaluate their own work from answer sheets. Teachers can encourage students to teach one another and, whenever possible, arrange for older children to give special help to children in lower grades. By their expression of attitudes, modeling, and nonverbal behavior, teachers communicate to learners that they are expected to be responsible and have the freedom in the classroom to demonstrate responsible behavior. It is possible that in developing the belief that one can be an originator, more is "caught" from teachers than is taught by them.

Using Feedback to Learn Mathematics

Informational feedback communicates to learners either that their goals have been attained because of their efforts or that certain improvements will enable them to reach their learning goals. It tells them *to think about their goals.*

Informational feedback is more useful to children than grades. Although grades from teachers communicate to children how well they have attained a standard, they do not tell learners the nature of their progress, nor do they emphasize student control of learning. Informational feedback tells children specifically what behavior shows achievement or what action will lead to attaining goals, and it conveys the idea that learning results from effort and leads to feelings of self-worth.

For example, if a child makes successful computer drawings with little trial and error, informational feedback tells her that she knows how to judge the length of a line and the number of degrees in a turn and that she should be pleased with what she has learned. However, if the child made many wrong estimates when drawing with a computer, informational feedback tells her to participate in special games or activities so that she can easily carry out her plans for creative drawings. She is also encouraged to engage in these games or activities to reach her goals, which will make her feel proud.

Informational feedback involves communication to children about learning. It tells them to tie effort to goals. It enables students to think about improving by their own endeavors. Teachers do not claim that learning mathematics is always easy, but they help students develop habits of persistence with assurances that goals can be reached with perseverance. Since the teacher has guided learners in setting goals that are attainable for them, success and motivation to achieve should be the outcome of instruction.

Informational feedback also tells learners that they can cope with learning mathematics and should, therefore, respect themselves. Such feedback focuses on the learner, not the teacher. The message it conveys is not "I, the teacher, want you to . . ." but rather "You, the learner, need to . . . in order to reach your goals, and attaining those goals will enhance your feelings of self-worth and pride."

CONCLUSION

As the title suggests, the key component of any instructional method is the means by which students can be motivated to learn mathematics. In the absence of such motivation, learning is reduced to a sequence of activities imposed by an agent external to the student, thereby leaving the student with a clear option of rejecting either the agent or the activities. Ultimately it is the student who must see and realize the joy and benefit of learning mathematics. Techniques were presented in this article for promoting such motivation: helping students internalize learning goals or reflecting on their own mathematical activity, for example. Facilitating self-motivation among students is, after all, one of our primary objectives in teaching mathematics.

REFERENCES

Ames, Russell, and Carole Ames. "Introduction." In *Research on Motivation in Education: Vol. 1. Student Motivation,* edited by Russell Ames and Carole Ames, pp. 1–11. New York: Academic Press, 1984.

Brown, Stephen I., and Marion I. Walters. *The Art of Problem Posing.* Philadelphia: Franklin Institute Press, 1983.

Cohen, Margaret. "Extrinsic Reinforcers and Intrinsic Motivation." In *Motivation Theory and Practice for Preservice Teachers,* edited by M. Kay Alderman and Margaret W. Cohen, pp. 6–15. Teacher Education Monograph No. 4. Washington, D.C.: ERIC Clearinghouse on Teacher Education, 1985.

DeCharms, Richard. "Motivation Enhancement in Educational Settings." In *Research on Motivation in Education: Vol. 1. Student Motivation,* edited by Russell Ames and Carole Ames, pp. 275–310. New York: Academic Press, 1984.

Deci, Edward, and Joseph Porac. "Cognitive Evaluation Theory and the Study of Human Motivation." In *The Hidden Cost of Reward: New Perspectives on the Psychology of Human Motivation,* edited by Mark R. Lepper and David Greene, pp. 149–76. Hillsdale, N.J.: Lawrence Erlbaum Associates, 1978.

Holmes, Emma E. *Children Learning Mathematics: A Cognitive Approach to Teaching.* Englewood Cliffs, N.J.: Prentice-Hall, 1985.

Holt, John. *How Children Fail.* Rev. ed. New York: Delacorte Press/Seymour Lawrence, 1982.

National Council of Teachers of Mathematics. *Curriculum and Evaluation Standards for School Mathematics.* Reston, Va.: The Council, 1989.

Resnick, Lauren B., and Wendy W. Ford. *The Psychology of Mathematics for Instruction.* Hillsdale, N.J.: Lawrence Erlbaum Associates, 1981.

Ryan, Richard M., James P. Connell, and Edward L. Deci. "A Motivational Analysis of Self-Determination and Self-Regulation in Education." In *Research on Motivation in Education: Vol. 2. The Classroom Milieu,* edited by Carole Ames and Russell Ames, pp. 13–51. New York: Academic Press, 1985.

Weiner, Bernard. "Principles for a Theory of Student Motivation and Their Application within an Attributional Framework." In *Research on Motivation in Education: Vol. 1. Student Motivation,* edited by Russell Ames and Carole Ames, pp. 15–38. New York: Academic Press, 1984.

Wittrock, Merlin C. "The Cognitive Movement in Instruction." *Educational Researcher* 8(1979): 5–10.

13

Assessment in Mathematics Classrooms, K–8

Norman Webb
Diane Briars

ASSESSMENT is the process of determining what students know. An active part of instruction in mathematics is checking what it is that students understand, getting feedback from students, and then using this information to guide the development of subsequent learning experiences. Continual assessment is particularly important in grades K–8, where students form the foundation for further learning in mathematics.

ASSESSMENT IS INTEGRAL TO INSTRUCTION

Because mathematics is a dynamic, interconnected system, students' knowledge of mathematical concepts and procedures, problem solving, and reasoning develop and mature over a period of years. The meaning that students assign to concepts and procedures in the early grades can greatly affect how they will perceive mathematics in high school and beyond. Just recording the number of exercises answered correctly, without understanding the student's thought and reasoning that went into producing those answers, will not be sufficient to know what meanings a student is assigning to the concepts and procedures that are being constructed. It is essential for effective teaching to know the meanings students are assigning to the mathematical ideas as they are learning to assure that a solid foundation is being formed. Assessment, then, must be an interaction between teacher and students, with the teacher continually seeking to understand what a student can do and how a student is able to do it and then using this information to guide instruction.

In this article we shall examine the assessment process and consider how assessment can provide the support necessary for effectively teaching mathematics. As a background to what follows, we take the view that assessment

(1) should occur in a variety of situations—for example, interview situations or class discussions, (2) should include a variety of mathematical representations, and (3) should involve the use of calculators and computers when appropriate. Such a variety of assessment methods intertwined with the instructional process is necessary in order to build upon students' conceptions of mathematics in the teaching of mathematics.

Getting Ready for Assessment

For many schools and school districts, the mathematics curriculum has been partitioned into very precise performance objectives. Here an objective can be an outcome a student can achieve with one or two days of instruction and practice or an outcome achieved over a longer period of time. Such objectives have been used to drive the assessment, which generally takes the form of a test or quiz containing a set of items with each set designed to measure a particular objective. To guide assessment that is interactive with instruction and that is to lead students to become mathematically powerful requires that student outcomes be stated more generally. The purpose of the outcome statement is not to provide all the details but to specify the connection among ideas and to help identify situations that can be used for instruction and assessment. For example, students in the early grades need to develop a strong understanding of different models for multiplication (repeated addition, combining equal groups, array, combinations, and area) so that they can continue to expand their knowledge to include factoring, proportions, percents, and polynomial expansion. A general outcome statement for guiding assessment and instruction of the concept of multiplication is the following:

> Students' conception of multiplication is to be developed through modeling, solving problems, and relating to real-world situations so that they can recognize problem structures that can be represented by multiplication, solve problems using a variety of multiplicative approaches, and multiply whole numbers.

One purpose of a goal statement, such as the one above, is to identify and plan assessment situations within instruction. If students are to use a variety of multiplicative approaches to solve problems, then one purpose for assessment is to determine if students in fact can solve problems in different ways. This means that students, after finding an answer in one way, need to be asked to find the answer using a different approach. Students also need to see that the same operation (here multiplication) can be used in solving a range of real-world problems, including combining equal groups, determining the number in an array, calculating "times as many," and finding the number of possible combinations, as in, for example, using two kinds of an item of clothing and three kinds of another item of clothing. This suggests that assessment needs to determine if students can identify a range of different multiplicative situations. One way of doing this is to ask students to generate different problems that can be solved using multiplication. If all the problems are of the same type—"Jody has five boxes of three golf balls each. How many golf balls does Jody have?"—then the student's knowledge of multiplicative situations needs to be explored further to determine if in fact the student's knowledge of such situations is indeed restricted to this one type or includes a variety of multiplicative situations.

Scoring and Reporting

Central to any assessment situation is scoring, recording, and reporting. For assessment that is integral to instruction, these procedures are usually done informally by giving instantaneous feedback to the student and not maintaining a formal record. However, there are times when it is necessary to record a student's progress or the progress of a class of students and make some report. Another reason for maintaining some record of students' progress is to reduce redundancy in instruction and assessment. On-going assessment will help to determine what students know in a variety of situations. If a student has already demonstrated an adequate knowledge of a concept or procedure during instruction and some record is kept, then assessing the student's knowledge on the same concept in the same way will not provide any new information and is not necessary. The record can be used to substantiate that fact. Another reason for recording some results from assessment, done as a part of instruction, is so that a variety of information can be collected. To draw meaning from this information, particularly when this has to be done for a large number of students, requires making some judgment about how all the information fits together and what conclusions can be made about each student's knowledge. A recorded trace of the student's assessment is helpful in doing this. This trace could be the accumulation in a file of marks on a checklist, brief comments on a note card, a list of scores, a sample of problems solved, and notes from other persons, such as other teachers or parents.

Scoring the work can vary from just a right/wrong rating to such variations as holistic and analytic scoring as described by Charles, Lester, and O'Daffer (1987). Both the holistic and the analytic scoring schemes assign multiple points (more than 0 and 1) to reflect the result obtained and how the result was obtained. Holistic scoring uses one scale to rate the total solution and answer. Analytic scoring breaks the work into parts, such as understanding a problem, the strategies used, and the answer obtained, and assigns a score to each part. A prior step to assessment is thinking through what scoring scheme to use, what records are needed, and how information is to be reported. An important consideration, however, is that these aspects of assessment do not become so overburdening that they detract from students' learning.

What to Assess

Determining the range of knowledge

Any assessment of students' mathematical knowledge should yield data about more than just students' computation skills; it should also include information about their knowledge of mathematics concepts and procedures and their problem-solving, reasoning, and communication skills (National Council of Teachers of Mathematics 1989). An assessment of a student's knowledge of concepts should provide information about whether the student discriminates between relevant and irrelevant attributes of a concept in selecting examples and nonexamples, is able to represent concepts in various ways, and recognizes their various meanings. An assessment of a student's knowledge of procedures should yield more than a measure of how reliably and efficiently the student applies procedures. It should also provide information on what a student knows about the concepts that underlie a procedure, when to apply procedures, why they work, and how to verify that these procedures give correct answers. An assessment of problem-solving skills should provide evidence of a student's ability to ask questions, use given information, make conjectures, derive solutions to problems, and generalize the solutions. An assessment of students' reasoning skills should provide evidence about different types of reasoning, including inductive and deductive reasoning, an analysis of situations, the development of plausible arguments, and the appreciation of the logical nature of mathematics. An assessment of a student's ability to communicate mathematically should be directed toward the expression of mathematical ideas, the understanding and interpretation of mathematical ideas, and the use of mathematical vocabulary and notation.

Specifying the content

Assessment helps to determine what a student knows about a topic or content area. To be sure that a student's knowledge of a topic is being fully assessed, some analysis of the range and depth of content knowledge is helpful. A good illustration of a content analysis for assessment is in the area of measurement. The results of an analysis, given in figure 13.1, list the expectations for what students should know and should be able to do. The list includes the integration of knowledge, more specific measurement skills, and the range of different types of attributes, units, and procedures that students are to know.

Integration of Knowledge
- Use measurements in solving problems
- Use measurements in other content areas besides mathematics
- Use measurement in solving situations from the student's experiences
- Use measures in doing mathematics involving numbers, computation, geometry, probability, statistics, patterns, and relationships

Measurement Skills
- Use appropriate units to reliably measure an attribute
- Recognize that all measurements are not exact and report measures as nearly, between, or about
- Estimate measurements and apply the estimate
- Compare measures of the same quantity using two or more different units
- Compare different quantities with the same unit
- Use an indirect measure such as speed and time to determine distance or proportions
- Produce or construct an object with attributes having given measures
- Use a variety of measurement strategies—iteration, comparison, direct match, formula, computation, simplifying (breaking into smaller parts)

Range of Measurement Concepts and Procedures
- *Attributes:* length, capacity, weight, area, time, temperature, and angle
- *Units:* inch, foot, yard, centimeter, meter, pounds, ounce, gram, kilogram, cubic inch, cubic foot, gallon, liter, quart, cubic centimeter, square inch, square foot, square meter, seconds, minute, hour, day, week, month, year, degree Fahrenheit, degree Celsius, degrees, and nonstandard units such as one hand, a finger, a crayon, and a matchstick
- *Procedure:* measuring process (identify a unit, compare with an attribute, and report the finding)

Fig. 13.1. Content analysis for developing an assessment within the area of measurement

The content analysis is a guide. The analysis provides a means of determining what range the assessment situations should span and what students should know. The analysis should not be used to define one assessment task to measure each aspect listed. Instead, situations should be used so that a number of different aspects of the measurement domain, listed in the analysis, can be assessed in one situation. The analysis then can be used to assure that all aspects of a topic are being assessed in at least some way, either during the normal progression of instruction or as a culmination to instruction.

A variety of situations can be used in assessing students' knowledge of measurement as the students are learning. One such situation for the primary grades is to have each student measure the length around the teacher's wrist. This provides the teacher the opportunity to observe closely what each student does, such as noting the beginning and ending points that are used and how students report the number of units. Drawing a graph of the results from all the students gives an occasion to explore what students know about the exactness of a measure and how measures can vary.

Another type of situation is to have students do a project like the following:

> Investigate how far you travel during a school day from the time you enter school in the morning to when you leave school in the afternoon.

This situation lets the teacher observe a number of strategies that students may use. In solving this situation, students can work in small groups, should have a variety of measuring devices available, and have access to calculators. As the students proceed, observations can be made of what units they select to use; how they use estimations; if they draw some form of scaled model of the school area; how they resolve measuring the distance traveled in a variety of locations, such as going up or down stairs or moving around while waiting in line; if they use indirect measures, such as noting that it takes two minutes to get to the playground . . . I go about 30 paces (45 feet) in 15 seconds so that means it is about 360 feet to the playground; and how they translate from one unit to another. This situation can be extended by graphing the distance traveled during the day by each student followed by comparing and explaining the differences.

In another situation the teacher draws figures of different shapes, including some irregular shapes, on grid paper. Without showing the students the figure, the teacher announces the area of a figure in square units. Students are to ask yes-and-no questions until they are able to draw the same figure on their own sheet of grid paper. The questions students ask will help determine their conceptions of area and other measures: "Does it have four sides?" "Are the corners square?" "Are the sides of the same length?" After so many questions, students should be able to draw the figure. Variations of this situation could include giving more information in the first clue, such as "I have a three-sided figure whose area is 10 square units" or having students work in small groups with one of the students having to guess the figure. Besides providing information on what students know about area and other attributes, this situation also lets the students communicate and receive ideas verbally.

These situations illustrate how students' knowledge of measurement can be assessed using relatively few situations that require integrating a range of concepts and procedures involving problem solving, reasoning, and communication. In planning such situations, the teacher will find that some form of an analysis is helpful for specifying the range and depth of knowledge that is to be assessed.

Selecting Tasks

Once the content to be assessed has been determined and situations identified, a next step is the selection of tasks for assessing specific knowledge. One criterion that should be used in this selection is that the task must evoke the knowledge that is to be assessed. Tasks for assessing a student's measurement skills should involve having the student measure something. Frequently items like the one in figure 13.2 are used to assess measuring skills. However, such tasks assess only the part of the process that involves the student reading scales.

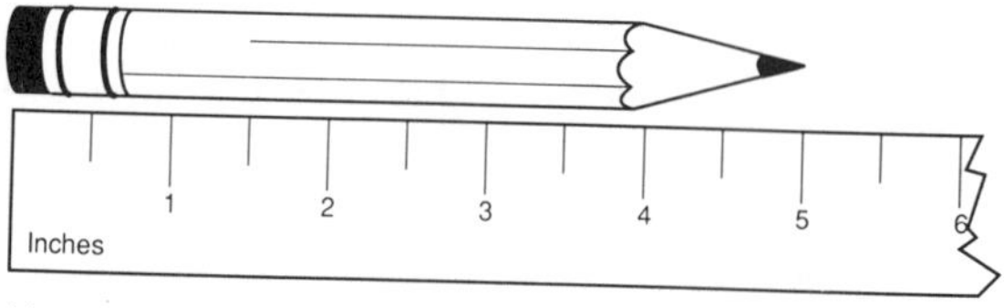

How long is the pencil?

Fig. 13.2. A poor task for assessing measurement

A more subtle consideration in deciding if a task evokes the knowledge being assessed is to determine whether a student can do a task correctly without having the knowledge to be assessed, that is, whether the student can get the right answer but for the wrong reason(s). For example, items like the one in figure 13.3 are often used to assess a knowledge of perimeter. However, a correct solution does not require any knowledge of perimeter: students can get a right answer by simply adding the dimensions of the figure, which is the most obvious thing to do when more than two numbers are given in a problem.

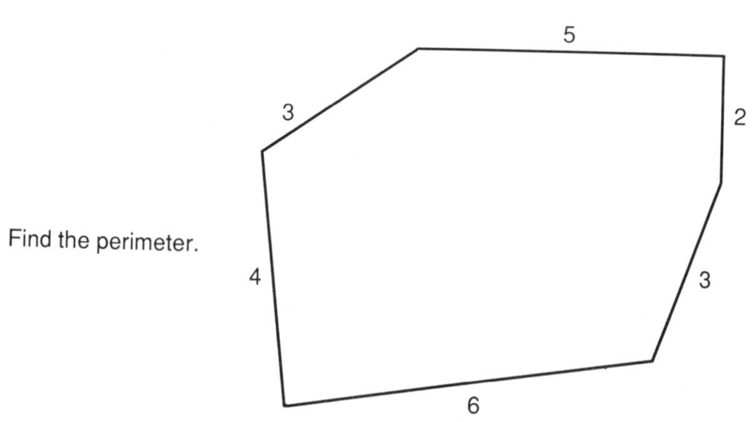

Fig. 13.3. A poor task for assessing a knowledge of perimeter

A second criterion for selecting tasks is to determine if the tasks will provide information about the extent of the student's knowledge being assessed. For example, an item like the one in figure 13.4 gives information about a student's range of conceptions of 1/2. The selection of (c) as an example of 1/2 suggests that the student does not recognize that halves must have equal areas. Selecting only (b) and (f) suggests that the student recognizes that 1/2 represents one of two equal pieces but does not recognize that the pieces do not have to be contiguous or congruent. An assessment task needs to be evaluated according to the amount of information the task provides.

Which figures show that exactly 1/2 of the region is shaded?

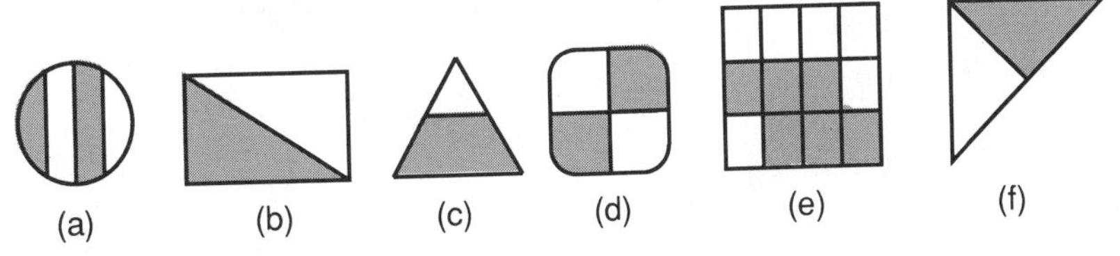

Fig. 13.4. One task for assessing different conceptions of 1/2

A third criterion is to select tasks that can give information about a udent's knowledge of a number of mathematical ideas and the extent to which the student has integrated them and is able to use them in new situations. For example, asking middle school students to administer a survey they developed and display the results in a circle graph provides information about a number of mathematical ideas. Developing the survey and reporting the results involve formulating a problem (e.g., identifying the variables, determining whom to survey, how many people to survey) and recording and organizing the data. Representing the results in a circle graph requires a knowledge of percent and angle measures, proportional reasoning (expressing data in percent, representing percents as sectors of a circle), and how to use a protractor.

Not all tasks can meet the last two criteria. However, within an assessment experience all the criteria should be met by at least some of the tasks.

ASSESSMENT AS INQUIRY

Assessment can be a form of inquiry where what a student knows is anticipated and different means of assessment are used to validate that initial hypothesis or to form a more accurate one. The Cognitively Guided Instruction (CGI) Project (Carpenter and Fennema 1988) is developing a model for classroom assessment that provides an example of such an inquiry approach. The CGI Project's guiding principle is that instructional decisions should be based on careful analyses of students' knowledge and learning goals. Consider an addition situation for a grade 1 student (Carpenter,

Fennema, and Peterson n.d.). In this example, assessment, in the form of an interview, provides information regarding a student's readiness to develop the count-on strategy. As the interview proceeds, the teacher bases the questions on the student's responses while keeping in mind the goal of determining if the student is ready to use the count-on strategy.

Teacher: There are six pennies in the bank. [The teacher places the pennies in the bank without counting each one.] How many pennies will be in the bank if we put in two more?

Paul: [Begins to count on his fingers to establish the number six.] One, two, three, four, five, six. [Hesitates and counts two more fingers; looks at the eight fingers.] Eight.

Teacher: Do you think you could solve the problem without counting all eight fingers?

Paul: [No response.]

Teacher: When you count, what number comes after six?

Paul: Seven comes after six.

Teacher: Right. Suppose we had seven pennies in the bank and we add one more penny. How many pennies would we have? Can you think of the number that is one more than seven when you count?

Paul: Well, . . . seven, eight. Eight comes after seven.

Teacher: Good. Let's put seven pennies in the bank. [Teacher places chips in groups.] If we put two more pennies in the bank, can we figure out how many pennies there will be altogether?

Paul: Seven, [pause] eight, nine. There are nine pennies.

By guiding Paul through the problem and getting feedback after each step, the teacher is able to ascertain that Paul seems to have the prerequisite knowledge for the count-on strategy. This can be confirmed by giving Paul a similar problem using the number fact $5 + 3$ to see if in fact he uses the strategy.

This dialogue depicts instruction as much as it represents assessment. The knowledge the teacher has of joining situations and of how one approach, counting all, can lead to a more advanced strategy of counting on is critical to the process. The teacher is able to develop a well-sequenced set of questions because of her or his understanding of what is being assessed, including determining what a student knows and the steps that can be taken for the student to develop a new strategy.

USING INFORMATION FROM ASSESSMENT

Assessment information can be used to determine students' perceptions of mathematical ideas and processes as well as their ability to function within a mathematical context. Through assessment, a better understanding should be obtained of how students are relating mathematical ideas to each other and if they are building an integrated notion of mathematics. Assessment information also is useful for making adjustments in instruction by helping to determine if another example is necessary or a different representation needs to be used. Making sure that assessment is integral to instruction should mean that the information obtained is directly useful for guiding instruction. In short, good assessment is good instruction.

REFERENCES

Carpenter, Thomas P., and Elizabeth Fennema. "Research and Cognitively Guided Instruction." In *Integrating Research on Teaching and Learning of Mathematics,* edited by Elizabeth Fennema, Thomas P. Carpenter, and Susan J. Lamon. Madison, Wis.: National Center for Research in Mathematical Sciences Education, Wisconsin Center for Education Research, 1988.

Carpenter, Thomas P., Elizabeth Fennema, and Penelope L. Peterson. "Assessing Children's Thinking." Working paper for the Cognitively Guided Instruction Project. Madison, Wis.: Wisconsin Center for Education Research, n.d.

Charles, Randall, Frank K. Lester, and Phares O'Daffer. *How to Evaluate Progress in Problem Solving*. Reston, Va.: National Council of Teachers of Mathematics, 1987.

National Council of Teachers of Mathematics. *Curriculum and Evaluation Standards for School Mathematics*. Reston, Va.: The Council, 1989.

14

Changes in Mathematics Teaching Call for Assessment Alternatives

David J. Clarke
Doug M. Clarke
Charles J. Lovitt

MANY things make up the complex picture evoked whenever one thinks of mathematics. Among these associations, it seems that one of the most enduring is the mathematics test. The identification of the subject mathematics with a single form of assessment has shaped student attitudes, teaching practices, and even the content of the subject. The preface to the NCTM *Curriculum and Evaluation Standards for School Mathematics* recognizes the importance of gathering valid information about student growth and achievement (National Council of Teachers of Mathematics 1989). The assessment alternatives that follow provide useful tools by which teachers might give meaning to such standards. Mathematics teachers are participating in a major restructuring of the goals and the practices of mathematics education. It is essential that assessment strategies be found that can adequately reflect this new conception of the subject—strategies that give recognition to the sorts of understandings that transcend individual topics and that will provide our pupils with tools of lasting value.

DEFINING ASSESSMENT

It is through our assessment that we communicate most clearly to students which activities and learning outcomes we value. It is important, therefore, that our assessment be comprehensive and give recognition to all valued learning experiences. Yet, frequently we assess to little purpose: collecting information we already possess, do not need, or on which we will never act. Further, in order to assess effectively within a mathematics curriculum that emphasizes applications and problem solving, we need assessment tools that are sensitive to process as well as product. The basis for many of these assessment strategies may be found in what we propose to call informal assessment. For our purposes, "informal assessment" is taken to be the collection of assessment information coincident with instruction. By contrast, "formal assessment" requires the organization of an "assessment event." It will be argued that informal assessment not only uses less class time but also commonly provides a better quality of information in a context in which the information can be put to immediate use.

When educational goals are restricted to the replication of mathematical procedures, conventional pencil-and-paper tests provide a picture of a student's level of performance. Since this assessment typically occurs at the completion of a course of study, information related to the learner's subsequent instruction in that topic is not a consideration. As our educational objectives broaden in scope, such assessment measures become increasingly inadequate. Instructional materials like the Geometric Supposers (Schwartz and Yerushalmy n.d.), through which students generate their own geometric theorems, provides learning experiences that would be grossly misrepresented by a subsequent pencil-and-paper test.

Assessment should do more than portray a learner's level of performance. It should guide the actions of all participants in the learning situation. If we accept the responsibility to direct our assessment toward guiding action, it is possible to state the major uses of assessment information in a way that anticipates subsequent action:

- To improve instruction by identifying the specific sources of a student's error that requires remediation or the specific learning behaviors that might need to be encouraged and developed or discouraged and replaced.
- To improve instruction by identifying those instructional strategies that are most successful.
- To inform the pupil of identified strengths and weaknesses both in knowledge and in learning strategies so that the most effective strategies might be applied where most needed.
- To inform subsequent teachers of the student's competencies so that they can more readily adapt their instruction to the student's needs.
- To inform parents of their child's progress so that they can give more effective support.

Desired learning behaviors will include skills and attributes that go beyond specific mathematical content, for instance, persistence, systematic working, efficient and effective organization, accuracy, conjecturing, modeling, creativity, and the ability to communicate ideas and procedures clearly. Although there is consensus on the importance of such learning objectives, they have seldom been the focus of assessment.

If assessment is to assume such an action orientation, then links must be forged among the assessment, the instruction it reviews, and the instruction it anticipates. Such a purpose will not be met through a test score alone; hence, it is essential that teachers expand their repertoire of assessment strategies.

In the remainder of this article, we discuss a range of alternative approaches to assessment, both informal and formal, as they have developed in Australian mathematics classrooms.

DOCUMENTING CLASSROOM OBSERVATIONS

Every teacher is continually offered a wealth of assessment information during the instructional process. Many act on this information, but few document it. It is through our documented assessment that we communicate most clearly to students which behaviors and learning outcomes we value.

It is clear that teachers formulate definite and quite accurate opinions concerning the competence of their pupils and that often formal assessment using tests does little more than legitimize and quantify the assessment made through extended classroom contact. However, informal assessment generally lacks structure, and the information it provides, although influencing teachers' decisions, is not systematically recorded and lacks the status accorded to a test score. Yet the quality of information collected informally is often higher than that obtained by conventional testing (see fig. 14.1).

> Consider a classroom activity exploring the scoring system used in Olympic Diving competitions. A student calculating the score for Kelly McCormick's final dive in the Los Angeles Olympics was confronted with the need to average 8.0,8.0,8.0,8.0,8.0. His teacher observed with interest that [8] [.] [0] was keyed in five times and the total divided by 5. The information gained by this observation is not just comparable to information provided by a related test item; *it is information of better quality, gained more efficiently, in a situation in which immediate constructive action is possible.*

Fig. 14.1

A teacher constructs a picture of each pupil's competencies through unplanned observations like the one depicted in figure 14.1, which are seldom systematically recorded. By introducing some structure into their observation, teachers can maximize the information they collect and minimize the time squandered on redundant, uninformative, and counterproductive assessment.

It is clearly impractical for a teacher to record comments on every pupil every day. Hence this strategy focuses only on recording *significant events.* A significant event is likely to be either

- atypical student behavior

or

- a clear illustration of new understandings or lack of understanding.

Make up a checklist of student behaviors, skills, or attitudes that you would like to foster. On a class list or in an exercise book record those significant moments that either challenge or extend your image of a student (fig. 14.2). Your original checklist will tell you what to look for and also provide a means to summarize the significant moments into a systematic record for each student. In identifying significant moments, ask yourself the question, "Will knowing this change my subsequent teaching of that student or that lesson?"

A related benefit arising from the use of such a procedure is that by the end of any given week there will be students for whom nothing was recorded. In this way, the use of annotated class lists may alert teachers to those "invisible" students present in every classroom.

Week beginning August 3	Annotated classlist	ACTION	
	COMMENTS (Aberrations and insights)	REQUIRED	TAKEN
Bastow, Barry	No concept of odd and even	*	
Carlton, Donna	Showed leadership in the group		
Carss, Marjorie			
Clements, Ken			
Caughey, Wendy			
Del Campo, Gina	Thought 63 and 36 the same	*	✓
Ganderton, Paul	Really tried		
Grace, Neville	Sequencing problems	*	
Howe, Peter			
Lee, Beth	Spatial thinker		
McDonough, Andrea	Recognised significance of a counter example		
McIntosh, Alistair			
Moule, Jim			
Mulligan, Joanne			
Nener, Kevin			
O'Connor, Fay			
Olssen, Kevin			
Palm[illegible]ael	M.A.B.[illegible]needed—[illegible]tens		
[illegible]			

Fig. 14.2

CREATING ASSESSMENT OPPORTUNITIES THROUGH QUESTIONING

All teachers engage in classroom questioning; it is one of the most fundamental teaching skills. It is through questioning that teachers and students establish a dialogue from which they all draw very specific conclusions regarding the relative competence of students. In addition, classroom questioning offers possibly the best chance to monitor the development of meaningful understanding (see, for example, Skemp [1976] and Davis [1978]). The openness of good questions can provide a powerful check on our teaching. Too often our examples limit students to a set of basic prototypes and do not allow them to explore the bounds and the structure of mathematical categories.

Non-Goal-Specific Tasks

It is possible to develop problem-solving expertise while still meeting the conventional goals of mathematical skills acquisition. For example, in figure 14.3 the conventional task, Find the side length x in the larger triangle, could be replaced by the less restrictive task, Find out everything you can about these two triangles. Obviously, the student who restricts the calculation to finding all angles and sides has responded at a different level from that of a student who not only calculates sides, angles, and areas but also draws conclusions about the ratio of areas when compared with the ratio of sides. "Problem solvers who calculate all that can be calculated from a set of givens can be expected to learn more than problem solvers who merely calculate values needed for solution" (Sweller, Mawer, and Ward 1983, p. 641). Such open-ended questions generate a variety of responses, all of which may be mathematically valid, differing only in the quality of understanding displayed.

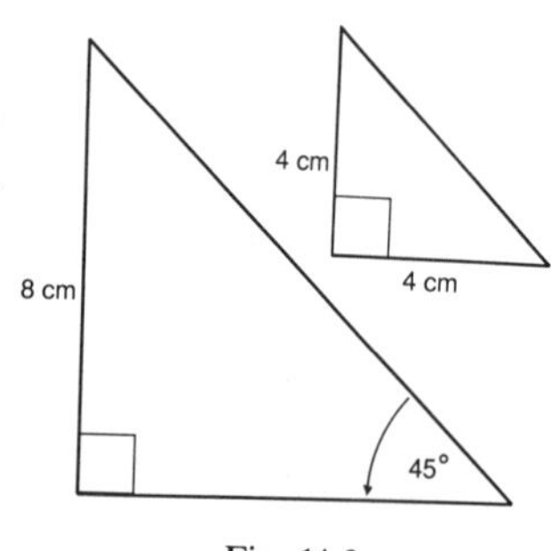

Fig. 14.3

A Five-Point Error Analysis

Consider this familiar scenario: The teacher has just delivered an introductory exposition masterly in its clarity and the richness of its examples. The class was attentive, interested, and involved. "Now I'd like you to try a few problems for yourselves." Several minutes of quiet industry ensued. One forlorn hand is raised.

"Please, Miss, I don't understand."

"Yes, Adam. What is it that you don't understand?"

"Everything."

To help Adam pinpoint his difficulty and to set the scene for some constructive assistance, the procedure in figure 14.4 is recommended.

"Find me one problem that you cannot do." (This should not be difficult!)

Procedure	*Purpose*
1. "Read the question to me. If you don't know a word, leave it out."	To identify reading errors
2. "Tell me what the question is asking you to do."	To identify errors in comprehension
3. "Tell me how you are going to find the answer."	To identify transformation errors
4. "Show me what to do to get the answer. Tell me what you are doing as you work."	To identify errors in the use of process skills
5. "Now write down the answer to the question."	To identify encoding errors

Fig. 14.4

Research reported by Clements (1980) indicated that at least 40 percent of students' errors on written mathematical problems occur before they even get to use the process skills their teachers have so laboriously stressed. In particular, students and teachers need to focus their attention on transforming a written task into a mathematical procedure. The five questions in figure 14.4 (Newman 1983) can usefully equip teachers and students to evaluate student understanding and identify the point at which difficulties occur.

MONITORING STUDENT SELF-ASSESSMENT

Effective assessment is a continuous process, predicated on the teacher's and the student's mutual recognition of the goals of the learning experience and the criteria for success. One of the most constructive and empowering educational goals we might frame would be to equip students to *monitor their own progress*.

The process of reflecting on one's learning is valuable in itself. Using a response sheet (Clarke 1987) gives children the opportunity to share their successes and concerns regularly with their teacher. The sheet, designed to be completed about every three weeks as a confidential communication from the student to the teacher, might contain questions like these:

- What is the most important thing you have learned in math this week?

- How do you feel in math class at the moment?
- What would you most like more help with?
- What is the biggest worry affecting your work in math right now?
- What one new problem can you now do?
- How could we improve math class?

Possible benefits to the teacher include acquiring a knowledge of students' difficulties with the content, a heightened awareness of prevailing student concerns, improved student-teacher rapport, and the identification of more appropriate, more effective means of instruction. Students may benefit through the recognition accorded to their concerns, the opportunity to reflect on and articulate their experiences with mathematics instruction, and the possibility that their instruction may become both more appropriate and more effective.

Student Work Folios

It is not only through questions that teachers gain insight into student understanding. One of the most effective ways to document student progress is to collect representative or significant samples of student work. How best should a teacher record the insights offered in the six-year-old's drawing of a clock shown in figure 14.5? To restrict one's documentation to a single mark would be to sacrifice precisely that detail that could most usefully guide our actions. A folio of student work yields a powerful illustration of a pupil's development, of particular relevance for report writing or parent-teacher interviews.

Fig. 14.5

Student Journals

In some schools, a central component of mathematical activity is the daily completion at home of a student journal. Through their journal-keeping activities, students are introduced to describing what they have learned, to summarizing key topics, and, progressively, to engaging in an internal dialogue through which they reflect on and explore the mathematics they have encountered. The assessment of student journals is accorded the same status as test performance or student performance in other assigned work. Teachers report formally on their assessment of student journals. Regular monitoring of the journals guides teaching practice and provides the basis for individual teacher-student discussion. Some excerpts from student journals (Waywood 1988, p. 7) are given in figure 14.6.

- Who on earth would have thought by looking at $q = 2.0 \times \frac{10^3}{1 \cdot 2} \div 10^4$ would equal $q = 1/6$ after it's worked out!! I certainly wouldn't have known!!!
- How come we class .9999 recurring as 1, when we don't consider .3333 recurring as a one-digit number or round it off?
- We were also told that pi is not a surd. I cannot agree with this. I believe pi is a surd because it has infinite non-recurring decimals. Why do you say then that it is not a surd????????
- You know it is really amazing but while I was trying to explain to Jo what it was I wanted her help with I ended up understanding it. Weird.

Fig. 14.6. Assessing by group report

The activity (Lovitt and Clarke 1988) shown in figure 14.7 illustrates the way in which graphs can tell a story. In small groups pupils are presented with a distance-time graph showing the movements of cars of different colors along a straight road. Each group writes a story or dialogue that relates to

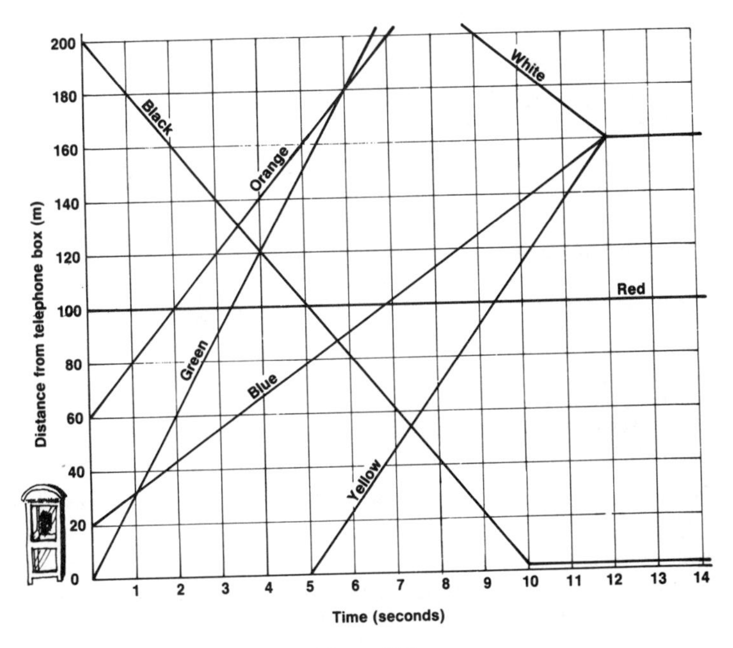

Figure 14.7

the experiences of the passengers of one of the cars. After preparing a story as one possible explanation for the graph, one pupil reads out the group's story while the others in the group act it out. As the reports are read, the rest of the class listens carefully and comments on the fidelity of the report to the graph, both in terms of consistency with the mathematical information and in terms of real-world plausibility.

A sample story

"We left the telephone box and jumped into our green Ferrari and raced off toward home at 36 meters per second (108 km/h)."

"That blue car twenty meters ahead of us is going pretty slowly!"

"*Time = 1.*" (intoned by the car microcomputer)

"As we passed the blue car (30 meters from the phone box) we could see an orange car up ahead, a red car parked by the side of the road, and 150 meters ahead there was this large black truck coming toward us."

"*Time = 2.*"

"Did you see that orange car nearly sideswipe the parked red car as it passed?"

"*Time = 3, and a bit.*"

"Hey, look! The red car belongs to Mr. Smith, our math teacher. He's got a flat tire (sympathetic mutterings) . . . ." [The saga continues.]

FINDING TEST ALTERNATIVES

So often, "tests measure how quickly people can solve relatively unimportant problems making as few errors as possible, rather than measuring how people grapple with relatively important problems, making as many productive errors as necessary with no time factor" (Blum 1978, p. 83). There are undoubtedly contexts and purposes for which the written test is wholly appropriate: for instance, in assessing the accurate replication of a basic procedure or the recall or recognition of factual knowledge. Although a written test may generally be an inappropriate measure of problem-solving techniques, particular "tool skills" have been identified, and these might well be the subject of a written test. It is imperative that we become more selective in our use of the written test. We must be convinced that it is the most effective strategy available to us, and that it will benefit a majority of our students. Alternative forms of testing should also be considered, such as the following.

Student-constructed Tests

Another approach to assessment is to administer a test made up entirely (or partly) of student-constructed test items. Not only do students have a new sense of participation in the assessment process, but there is a far greater interest in discussing the solutions to their test. The preparation of test items also is a most effective revision strategy. We have found the following method to be successful:

- Divide the class into groups of two or three students and ask each group to make up five problems that they feel would test the content (unit or topic, concepts or skills) fairly.
- As teacher, you reserve the right to select from the pool of problems generated in compiling the test, but a guarantee to include at least one item from each group gives legitimacy to the exercise in the minds of the pupils.
- Your editorial rights include the right to rephrase an ambiguous or poorly worded question while remaining true to the essential form of the students' question.
- In reviewing class performance on the test, you will find it useful to ask the group responsible for a particular question to outline the sort of solution they were after.

Practical Tests

Many topics lend themselves to a practical test—for instance, topics from measurement, pattern and order, geometry, and probability and statistics.

Consider the following topic of volume, adapted from the British Assessment of Performance Unit reported in Foxman and Mitchell (1983, p. 3).

Practical test—Volume

Materials required: Solid wooden cube, solid metal cylinder, stone, ruler, water, measuring cylinder, displacement can, paper, pencil

Task: Use the equipment provided to find the volume of the cube, the cylinder, and the stone.

A variation of this technique is to test computing skills in a practical situation. One approach successfully used by a secondary school teacher involved distributing a weekly list of computing skills to be acquired. During a particular week, each student would be asked to demonstrate one of the skills listed for the previous week, the particular skill being chosen by the teacher without prior notice to the student. Such a sampling approach to the assessment of practical skills is just as rigorous as the conventional test, whose questions represent a "lucky dip" of the course content.

Added benefits of practical tests are the provision of short-term learning goals, enhanced motivation, immediate and unambiguous feedback, and a high degree of assessment validity, since the skills are assessed in practice, as they were learned and as they will be applied.

CONCLUSION

The approach taken in this article is predicated on certain beliefs regarding the function of assessment in mathematics:

1. *It is assessment that enables us to distinguish between teaching and learning*. Beginning teachers often hold the naive view that what was taught will correspond to what was learned. One of the hard lessons of professional growth as a teacher comes with the recognition of this essential distinction and the responsibility it places on teachers to monitor regularly each pupil's learning.

2. *Assessment should anticipate action* and is best able to do so when it is coincident with instruction. Assessment should guide the actions of all participants in the learning situation. If we are to realize this action orientation, links must be forged among the assessment, the instruction it reviews, and the instruction it anticipates.

3. *It is through our assessment that we communicate most clearly to students those activities and learning outcomes that we value*. Many of the assessment strategies required are already in use and need only to be systematically applied and documented to enable us to give all our educational goals the recognition of appropriate assessment.

The issues raised in this article and the techniques outlined are a direct result of the Australian Mathematics Curriculum and Teaching Program (MCTP), which sought to document the wisdom of practice as evidenced in mathematics classrooms across Australia. The results of this national project can be found in a resource collection that documents the best examples of effective mathematics teaching, professional development, and assessment alternatives from Australian mathematics classrooms (Clarke 1989).

As teachers adopt a range of new approaches to the teaching and learning of mathematics, there is growing consensus that traditional paper-and-pencil tests are inadequate in providing useful assessment information. The range of assessment alternatives outlined in this article are fully supported by the NCTM's *Curriculum and Evaluation Standards* (NCTM 1989), and teachers are invited to try them with their students and gauge their effectiveness in supporting the new goals of mathematics education. We have a professional obligation to ensure that our assessment contributes constructively to the learning of our pupils and to ensure that all those skills and attributes that we most wish to foster receive the recognition of appropriate assessment.

REFERENCES

Blum, Jeffrey M. *Pseudoscience and Mental Ability.* New York: Monthly Review Press, 1978.

Clarke, David J. "The Interactive Monitoring of Children's Learning in Mathematics." *For the Learning of Mathematics* 7 (1987): 2–6.

———. *MCTP Professional Development Package: Assessment Alternatives in Mathematics.* Canberra, Australia: Curriculum Development Centre, 1989.

Clements, M. A. (Ken). "Analyzing Students' Errors on Written Mathematical Tasks." *Educational Studies in Mathematics* 11 (1980): 1–21.

Davis, Edward J. "A Model for Understanding Understanding in Mathematics." *Arithmetic Teacher* 26 (September 1978): 13–17.

Foxman, Derek, and Peter Mitchell. "Assessing Mathematics: APU Framework and Modes of Assessment." *Mathematics in Schools* (November 1983): 2–5.

Lovitt, Charles J., and Doug M. Clarke. *MCTP Professional Development Package: Activity Bank Volume 1.* Canberra, Australia: Curriculum Development Centre, 1988.

National Council of Teachers of Mathematics. *Curriculum and Evaluation Standards for School Mathematics.* Reston, Va.: The Council, 1989.

Newman, Anne. *The Newman Language of Mathematics Kit.* Sydney: Harcourt, Brace & Jovanovich, 1983.

Schwartz, Judah, and Michal Yerushalmy. The Geometric Supposers. Cambridge, Mass.: Education Development Center. (Available from Sunburst Communications, Pleasantville, N.Y.)

Skemp, Richard R. "Relational Understanding and Instrumental Understanding." *Mathematics Teaching* 76 (1976): 20–26.

Sweller, John, Robert F. Mawer, and Mark R. Ward. "Development of Expertise in Mathematical Problem Solving." *Journal of Experimental Psychology: General* 112 (1983): 639–61.

Waywood, Andrew. "Number and Meaning." Mimeographed. Richmond, Victoria, Australia: Vaucluse College, 1988.

15

Mathematics for All Americans

Lynn Arthur Steen

A GOOD teacher is one who stimulates students to learn. Thus sound policy for mathematics education must look as much at the students being educated as at the mathematics being taught.

Sound policy concerning economic competitiveness also leads directly to a focus on trained personnel. People, not facilities or equipment, are the ultimate source of technological innovation. Yet the creation of a scientist or engineer takes nearly two decades of education grounded in the study of mathematics.

Today's students—tomorrow's scientists—are not the same as yesterday's students, and tomorrow's students will be even more varied. Demographic trends alter the nature of mathematics education as surely as any other force for change. As powerful as they are inexorable, these trends are too often hidden from public view. Nonetheless, the new demography will forever change the nature of America.

DEMOGRAPHIC TRENDS

Several issues make demography especially important for mathematics education. First, minorities are becoming a majority (Oaxaca and Reynolds 1988). In twenty-three of the nation's twenty-five largest public school systems, including each one of the top ten, total enrollment of minorities is over 50 percent. By the year 2000, 40 percent of the children in public schools will be black or Hispanic, compared to 13 percent at the end of World War II.

The study of mathematics in the United States, however, has been dominated by white males of European descent (and as often as not, European educated as well). Only 4 percent of the bachelor's degrees in mathematics and fewer than 2 percent of the Ph.D. degrees go to U.S. blacks or Hispanics, our largest minority populations. Three of every four U.S. citizens who receive Ph.D. degrees in mathematics are not only white, but male.

Yet according to a recent projection of the U.S. work force in the year

2000 (Johnston and Packer 1987), only 15 percent of the young people entering the labor force around the turn of the century will be white males. Moreover, the fraction of new jobs requiring a full four years of high school mathematics is expected to be 60 percent higher in the late 1990s than in the 1970s ("Human Capital" 1988).

Another demographic phenomenon that bears heavily on mathematics education is the effect of population waves on the supply of mathematics teachers (Hodgkinson 1985). The trough in the post-postwar population wave is just now passing through the schools, so total school enrollments are beginning to climb after a fifteen-year period of decline. However, the number of twenty-two-year-olds is entering a long period of decline, from a high of nearly 4.5 million in the early 1980s to a low of about 3.5 million in 2000.

This means that the pool of potential new teachers will decline just as school enrollments begin to build. Pressure in mathematics education will be especially severe because demand for mathematics instruction is rising as a consequence of the increasing use of mathematics in business and technology. Yet this same increase in nonacademic uses of mathematics is also providing many attractive job alternatives for those who are prepared to teach high school or college mathematics.

So at a time when the nation is facing a crucial shortage of mathematicians—when, for example, the National Security Agency alone is trying to hire as many U.S. mathematics Ph.D.'s as are educated each year—we find ourselves in the untenable position of recruiting virtually all our future mathematicians from only 15 percent of the nation's new workers.

MATHEMATICS PIPELINE

One in four Americans fails to finish high school on schedule. One of the three who do finish on time will graduate marginally literate and virtually innumerate. The other two will pursue some form of higher education. With few exceptions, most students who enter the work force directly out of high school—with or without a diploma—are barely able to hold marginal jobs and are totally lacking in skills needed for advancement or professional development.

According to a recent literacy study by the National Assessment of Educational Progress (Kirsch and Jungeblut 1986), fewer than 40 percent of young adults can carry out simple restaurant calculations (adding the cost of two items, adding a tip, and determining change). Those who can do these types of homely mathematical tasks are the students who enter postsecondary institutions. Those who can't move directly into business, industry, and the military, where remediation programs for workers cost almost as much as what is spent annually on public school education.

The pipeline of mathematics students that flows from kindergarten to graduate school is more like a refinery than a simple pipeline: open valves here and there drain away valuable talent; clogged filters in certain parts prevent normal flow; and feedback loops at crucial junctions permit students to become teachers while still also being students. The entire system is extraordinarily complex, hence very difficult to regulate.

Apart from English, mathematics is the most dominant, most expensive, and most influential constituent of our schools, involving 25 million schoolchildren, 10 million secondary school students, and 3 million college students. Mathematics courses account for 20 percent of all school instruction and for 10 percent of all course credits in higher education. Mathematics represents nearly two-thirds of total precollege instructional effort devoted to science; even in higher education, mathematics credits account for nearly one-third of the total devoted to science and engineering.

The dropout rate from open valves in the mathematics refinery is staggering. From ninth grade through the Ph.D., the half-life of students in the mathematics curriculum is one year: beginning with approximately 3.2 million students entering high school, we lose 50 percent each year until only a few hundred attain the Ph.D. On a hand calculator divide 3.2 million repeatedly by 2: you will witness the annual productivity decline of mathematics education in the United States.

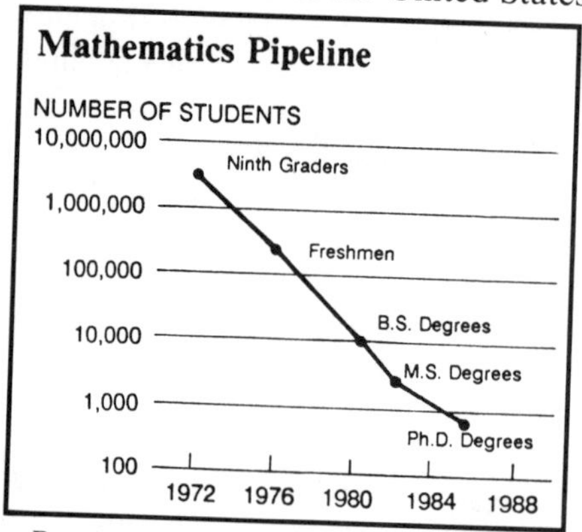

Reprinted by permission from page 6 of *Everybody Counts: A Report to the Nation on the Future of Mathematics Education,* Washington, D.C.: National Research Council, National Academy Press, 1989.

Losses from the mathematics pipeline come disproportionately from females, blacks, Hispanics, and Native Americans, but at very different stages (National Science Foundation 1989). Blacks drop out early, at twice the national average. Hispanic dropout rates are not as high but are much worse than the national average, which itself is unacceptably high. Women persist through college at rates comparable to men, but then drop out at much higher rates in graduate school.

One-fourth of our population—the underrepresented minorities—currently produce fewer than 2 percent of our scientists, mathematicians, or engineers. Future indicators suggest little change in this pattern. For example, virtually none of the black college freshmen who score highest on the SAT mathematics test indicate interest in majoring in mathematics.

These facts underscore a demographic reality that will shape mathematics education for the rest of this century: the fastest-growing segments of the American population are those that are most likely to drop out of the mathematics pipeline.

Consequences

As science and society have become more mathematical, demand for mathematics in colleges has doubled. Yet because so many students who need further mathematics enter college weak in mathematics, high demand has significantly diminished the quality of undergraduate mathematics. Indeed, on a national scale, undergraduate mathematics has been reduced from higher education to high school education.

Approximately one-fourth of the mathematics enrollments in colleges and universities are in strictly remedial courses, ranging from arithmetic to high school algebra II. Nearly half of the enrollments are in advanced high school–level mathematics that does not require sophisticated tools like calculus. Only about 30 percent of college enrollments are in traditional college-level mathematics—25 percent in calculus and 5 percent in more advanced courses.

A more distant consequence of this decline in the average level of undergraduate mathematics instruction is that U.S. students can no longer compete effectively in our own mathematics graduate schools. Fewer than half of the Ph.D. degrees in mathematics awarded by U.S. graduate schools now go to U.S. citizens.

These diverse data from different locations in the mathematics refinery underscore the complexity and fragility of the system. To reduce the dropout rate will require energetic and well-trained teachers, especially from underrepresented minorities. But to find such teachers, we need to reinvigorate the source of mathematics teachers: undergraduate mathematics. To do *that* will require an infusion of new Ph.D.'s to replace the imminent wave of retirements from the postwar generation of faculty. This brings us back to where we began—to the need for more students in the mathematics pipeline so that there can some day be more well-educated schoolteachers as well as more university professors.

No single response is sufficient to the challenge we face. Unless we improve the attractiveness of mathematics at *all* levels of schooling and for *all* socioeconomic groups, we shall never be able to attract enough young people into science and engineering careers to retain our competitive position into the next century. It does no good to bring college biology into the age of genetic engineering if potential biology students dropped out of high school algebra before ever getting to college. We must first turn off leaks in the refinery and unclog its filters. Most important, we need to understand the refinery as a whole before we try to adjust any particular part.

Recommendations

The refinery of mathematics education produces many different products from the same raw material: schoolteachers, scientists, programmers, economists, technicians, college professors are siphoned off a various points in the educational process. The performance of this system is of vital importance to our nation: adequate outputs at different levels must be assured if we are to maintain a strong economy and effective national security.

Significant change must be built on public support (National Research Council 1989). The first steps in bringing about change must be to convince the public of certain realities:

- That mathematics is the foundation discipline for science and technology
- That far too many minority children leave school without having acquired the mathematical power necessary for productive lives
- That all children—not only those with special talents—can learn mathematics
- That our children must learn a different kind of mathematics for the future from what was adequate in the past
- That confidence rather than calculation should be a chief objective of school mathematics
- That our nation's economic future depends on strength in mathematics education

Most important, we need to stimulate in *all* Americans a sustained demand for the type of school mathematics that will be required by those who live and work in the twenty-first century.

REFERENCES

Hodgkinson, Harold L. *All One System: Demographics of Education—Kindergarten through Graduate School.* Washington, D.C.: Institute for Educational Leadership, 1985.

"Human Capital: The Decline of America's Work Force." *Business Week,* Special Report, 19 September 1988, pp. 100–141.

Johnston, William B., and Arnold E. Packer, eds. *Workforce 2000: Work and Workers for the Twenty-first Century.* Indianapolis, Ind.: Hudson Institute, 1987.

Kirsch, Irwin S., and Ann Jungeblut. *Literacy Profiles of America's Young Adults.* Princeton, N.J.: Educational Testing Service, 1986.

National Research Council. *Everybody Counts: A Report to the Nation on the Future of Mathematics Education.* Washington, D.C.: National Academy Press, 1989.

National Science Foundation. *Women and Minorities in Science and Engineering.* Washington, D.C.: The Foundation, 1988.

Oaxaca, Jaime, and Ann W. Reynolds. *Changing America: The New Face of Science and Engineering.* Interim Report. Washington, D.C.: Task Force on Women, Minorities, and the Handicapped in Science and Technology, 1988.

16

The Challenges of a Changing World for Mathematics Education

Walter G. Secada

CHANGING demographics and changing economic and social orders in the United States constitute forces that create a mixture of hope and crisis in mathematics education. How these forces interact to create that mix and how mathematics educators might respond to meet the challenge are examined in the following sections of this article.

CHANGING DEMOGRAPHICS

In 1976, 24 percent of the total student enrollment in U.S. schools was nonwhite; by 1984, the figure had risen to 29 percent; by the year 2000, close to 40 percent of the country's school population will be minority (Center for Education Statistics [CES] 1987a, p. 64; Hodgkinson 1985). In the country's twenty largest school districts, the respective figures for 1976 and 1984 were 60 percent and 70 percent (CES 1987a, p. 64). In New Mexico, a plurality of the school-aged population is Hispanic. The year 1988 marked the watershed for California; it was the year in which nonwhite minorities became the majority population. In Texas and other states, the majority-minority split is, for all practical purposes, 50-50 (*Resegregation of Public Schools* 1989).

An earlier version of this article was presented at the Technical Assistance Institute on Project Management and Evaluation, sponsored by the Office of Bilingual Education and Minority Language Affairs (OBEMLA), U.S. Department of Education, Washington, D.C., 29 February–2 March 1988. The preparation of this article was supported in part by OBEMLA and the Wisconsin Center for Education Research (WCER), School of Education, University of Wisconsin—Madison. Opinions expressed are those of the author and do not reflect the positions of either OBEMLA or WCER.

One in four students is poor (Kennedy, Jung, and Orland 1986, p. 71), and one in five students lives in a single-parent home (CES 1987b, p. 21). Not only will these rates increase, but also students will share more than just one characteristic. For example, children from single-parent homes are more likely to be poor than those from homes with two parents. Black and Hispanic children are more likely to be poor than whites. And a black child who is poor is likely to spend most of his or her childhood in poverty, whereas a white child is likely to get out of poverty in fewer than five years (Kennedy, Jung, and Orland 1986).

Among children from minority-language backgrounds, increasing numbers are entering school with little or no competence in the English language (Hispanic Policy Development Project 1988; O'Malley 1981). Though Spanish is the predominant first language for these children—and it is likely to remain so into the next century (Veltman 1988)—increasing numbers of children with limited English proficiency (LEP) are entering school with non–Spanish-language backgrounds, such as Arabic, Chinese, Hmong, Khmer, Lao, Thai, and Vietnamese (Oxford-Carpenter et al. 1984).

Hispanics are the fastest growing group in the United States. From 1982 to 1985, the Hispanic population grew an average of 3.0 percent a year, the black population grew 1.6 percent a year, and the white, non-Hispanic population grew by 0.6 percent. Hispanics are projected to become the largest minority group in the United States around the year 2010 (U.S. Bureau of the Census 1986). Though Hispanics have shifted from rural to urban population centers and are found throughout the United States, over 60 percent of them live in California, New York, and Texas. Adding Florida and Illinois raises the total to 75 percent (Arias 1986, p. 29).

The social structure of educational opportunity for minorities is highly stratified. Increasingly, Hispanics are being segregated in schools (Arias 1986; Espinosa and Ochoa 1986). And blacks who attend nominally desegregated schools are tracked or grouped in other ways that reduce their opportunities to learn mathematics (*Resegregation of Public Schools* 1989).

Hence, not only do correlations exist among those demographic characteristics of poverty, family structure, and ethnicity for which indicators of participation in mathematics are distressingly low, but also the projected growth in our school-aged population is precisely for those groups for whom the school system has not worked as well as it might.

The changing demographics are affecting not only the school-aged population but also the composition of the U.S. work force. The entry level work force is shrinking in absolute terms while minorities are increasing their numbers. These changes in the nature of the American work force furnish some hope for employment opportunities for women and minority students. Of the net number of newly created jobs in 1985–2000, a scant 15 percent are projected to be filled by white males. White females will fill 42 percent of these new jobs; native, non-white males and females will fill 7 percent and 13 percent, respectively; immigrant males and females, the remaining 13 percent and 9 percent (Johnston and Packer 1987, p. xxi). In the report of a national leadership meeting, *Youth 2000,* the National Alliance for Business (1986b) articulates how the delicate balance between challenge and opportunity may be irretrievably lost and hence take on crisis proportions. Thomas D. Bell, president and chief executive officer of the Hudson Institute, noted:

> We have a unique opportunity. But, it's fleeting. The problem of youth unemployment can be solved by the turn of the century. That is something we have never been able to say before, and perhaps will never be able to say again. (p. 15)

Admiral James D. Watkins, Chief of Naval Operations, concurred:

> There is a confluence of national thought right now, there's an opportunity that presents itself to all of us, and if we don't grasp the opportunity, we're going to plunge the nation into economic and military crisis. By the end of this decade, the military will require one out of every two qualified males. Industry, business, will all be clamoring with us [the military] for the same resources that simply are not going to be there in the kinds of numbers this nation needs. (p. 23)

The opportunity arises from the shrinking size of the entry labor force tied to an increasing demand for labor. In this opportunity lies the promise of full employment, provided that we meet the challenge of an educated labor pool. The danger lies in the missed opportunity, and it has multiple facets. First is the creation of a pool of people who will be illiterate and will remain a permanently unemployable underclass, regardless of industry's and the military's need for workers. They represent an economic drain on our social, welfare, and justice systems and a political danger to the country's well being. Second, if supply does not meet the demand for well-educated workers, then the United States will lose its economic and military prestige and position as it watches other, more prepared nations take advantage of the same opportunity. A third danger is implicit in those projections. This last danger concerns the overall ability and willingness of today's youth to support an aging society that failed to educate them in the first place, thereby condemning them to a life of second-class power, economically and militarily, both as a nation and individually. In response to the second danger, the National Alliance of Business (1986a) has called for a review of U.S. immigration policy so that industry can meet its demand for skilled workers from abroad. There is no ready response to the first and third dangers.

THE CHANGING ECONOMIC AND SOCIAL ORDER

Not only is the country's school-aged population changing, but so is the world for which schools are preparing these students. The level of mathematical literacy needed to participate fully in that world, its jobs, its economic and social orders, and its democratic institutions is steadily increasing.

New jobs will be primarily in the service sectors of the economy. The fastest-growing job areas require highly skilled workers: natural scientists, lawyers, engineers, managers, teachers, technicians (Johnston and Packer 1987, p. xxii). The small number of qualified people who will be available for these new jobs and the increased skill levels they require support the view that we must more fully educate more women and minorities, if for no other reason than enlightened self-interest.

Today's jobs filled by minorities tend to be in manufacturing or agriculture. These require lower levels of mathematical and scientific literacy than jobs in the developing sectors of the economy (Johnston and Packer 1987). Yet even these low-level jobs are being transformed by computers and other technological advances into the workplace. Finally, the armed forces are competing with industry for people with these higher levels of technologically based literacy (National Alliance of Business 1986a, 1986b).

Poverty, jobs, and economics are but part of the social whole. The mathematical literacy necessary to gain access to the most basic services might be very low, but an understanding of how those services are allocated, and hence the potential for gaining *control* of them, requires much higher levels of literacy. For example, buying food in a market is easy. Understanding the forces that lead to higher food prices in specific neighborhoods and deciding either to buy elsewhere or to challenge the assumptions that underlie those forces are much more difficult. One can think of various situations in which low levels of mathematical literacy do not rule out a particular service: just go to a service broker. But such low literacy rules out the possibility of an individual's gaining control of that service.

Grubb (1986) argues that the demands for increased mathematical and scientific literacy should not be constrained by the overly narrow link to employment and economics that reports such as *A Nation at Risk* (National Commission on Excellence in Education 1983) have tried to forge. He claims that the "variety and complexity of scientific issues in the *political* realm will surely increase, [hence] the prerequisites for informed *citizenship* now include the basics of math and science" (p. 27, emphasis added). Without such knowledge, the citizenry will be unable to make informed decisions, and our democratic institutions will be threatened. Strike (1985), *A Nation at Risk,* and Grubb (1986, p. 27) all quote Thomas Jefferson on this point:

> I know no safe depository of the ultimate powers of the society but the people themselves; and if we think them not enlightened enough to exercise their control with a wholesome discretion, the remedy is not to take it away from them, but to inform their discretion.

The American democratic process has become more complex because of advances in science and technology. We are confronted with choices whose complexity in moral as well as technical terms demand a mathematically literate populace. People without such literacy will find themselves abdicating their roles in our most cherished democratic institutions.

MEETING THE CHALLENGE

What should mathematics educators do? As a general policy, we should remove the "computational gate" to the study of mathematics in high school. In so doing, we should provide all students with the opportunity to study the more advanced forms of mathematics that will be critical for their participation in our society. The *Curriculum and Evaluation Standards for School Mathematics* (National Council of Teachers of Mathematics 1989) provides one possible framework in which to do so.

In schools, we should attend to how students are being resegregated and how their opportunities to learn mathematics are thereby being limited. We should be concerned with how we structure our classrooms, how we communicate with parents and policy makers, and how our own professional growth and development help us address these larger social issues as well as those that help us improve the practice of our craft.

For minority students, most roads lead to general or remedial mathematics as the terminal mathematics experience in high school (Oakes 1987). Given that experienced mathematics teachers often compete to teach the more advanced classes, an imbalance exists between the expertise of the teachers available to teach the students and the expertise needed for that population of students. It is not uncommon that beginning and less senior teachers or those who have fallen out of favor with the administration are assigned to teach general mathematics. This practice of assigning teachers results in those classes of students who need the highest quality of instruction being taught by the least experienced teachers. Also, this practice sends some very powerful messages to both teachers and their students about career ladders, the rewards of seniority, and what it means to teach minority students who tend to be overrepresented in low-level classes. Courses could be assigned equally among a school's faculty. Alternatively, more experienced mathematics teachers might be encouraged to teach remedial courses and be rewarded for doing so. Moreover, as the *Curriculum and Evaluation Standards* suggests, remedial and general mathematics courses might be restructured to make them less deadly for both teachers and students and to change them from being terminal courses to those that lead to additional studies.

Course taking is the single most powerful factor under the schools' control that affects student academic achievement (Dossey et al. 1988; Myers and Milne 1988; Oakes 1987). Yet, minority students do not receive academic counseling encouraging them to take advanced courses (MacCorquodale 1988; Oakes 1987). We should recognize that minority populations have traditional values that, to say the least, stress gender-differentiated roles for boys and girls. Hence, minority girls are likely to be at a double disadvantage when being encouraged to take mathematics courses. Mathematics teachers and counselors need to be alert to the social forces that affect their students' decisions to take courses. And they need to communicate to their students the importance of course taking for future careers, participation in society, and, frankly, earnings.

Access to computers is typically limited to students who are taking advanced mathematics courses (Oakes 1987), but many software applications do not require extensive mathematical skills. In fact, they might serve to develop precisely those advanced skills. For example, spreadsheets can be used to model some rather sophisticated processes in business as well as in science. Certainly by middle school if not before, all students should have ready access to calculators, and the focus of the curriculum should be enlarged beyond number and computation.

Schools and mathematics teachers need to anticipate the changes that are coming in the mathematics education reform movement. They should get in on the ground floor and insist that those changes include attention to equity. We simply cannot afford to ignore educational reform and let the education of females and minorities, once again, play catch-up to a moving target. For example, the *Curriculum and Evaluation Standards* argues that in elementary school, mathematics should be taught in a developmentally appropriate manner. Manipulatives and contextually rich problem situations are two areas in which future curricula should be strengthened. We know from the Significant Bilingual Instructional Features Study (see Tikunoff 1985) that cultural referents should be included in the instruction of LEP students. Hence, the newly developed mathematics curricula should include multicultural referents, and they should create problem-solving situations that are understandable by children from diverse cultural backgrounds.

Finally, in-service staff development must take account of the forces that are changing our world. It is not enough to provide teachers with workshops on the *Curriculum and Evaluation Standards;* they need help in implementing those standards so that *all* children can participate. Moreover, the content and tone of such efforts needs to change.

After more than thirty years since the first Sputnik-driven reforms, curriculum development is finally taking into account how children think. It should not take another thirty years for teacher education and staff development to come to a similar realization. Mathematics teachers make numerous decisions about curriculum, classroom organization, and the modification of their instructional plans to facilitate learning (Clark and Peterson 1986). Yet teachers also make decisions and engage in behaviors that may constrain student opportunities to learn mathematics (Oakes 1987). The challenge for teacher-development efforts is not only to respect the autonomy and thinking of mathematics teachers but also to urge them to examine the effects of their own behaviors and attitudes on the education of minorities. Moreover, teacher-development efforts need to help teachers understand how they can succeed.

CHALLENGES, NOT EXCUSES

The fact of change—demographic, social, and curricular—poses a challenge to mathematics educators. If we are to meet this challenge, we should keep in mind the distinction between a challenge and an excuse for the failure to even try. For example, Orr (1987a) provides a comprehensive analysis of how black students experience difficulties in solving high school algebra and geometry problems. She traces those difficulties to the students' failure to discriminate the fine mathematical distinctions that are essential for understanding the texts they are reading. She argues that this failure is intimately related to the fact that Black English Vernacular (BEV), which her students speak, does not include the prepositional structures of standard English. That message—that something is wrong with BEV—is the one the media have picked up (Orr 1987b, 1987c).

Unfortunately, the press has failed to note the efforts of Orr and her staff to meet the challenges posed by those students who enter her school's mathematics classes using BEV. Orr's efforts are based on accepting these students as they are and then moving them into the standard curriculum. Neither she nor her teachers have lowered their expectations of their students because of their purported deficits.

Orr (1987a, pp. 201–2) writes about the compromises that schools and mathematics teachers have been making in the face of the challenges posed by black students:

> I hear more and more about situations where teachers, understandably discouraged with high failure rates, gradually, and more often than not unconsciously, modify what they do and require. It is painful to face unremitting large numbers of students who do not understand and not be able to get them to understand; good teachers blame themselves and try again and again. The current pressures are, perhaps understandably, resulting in [a] focus on the memorizable, the replicable: in math, computation becomes the first priority.

Orr (1987a, pp. 206–8) also describes how she and her staff have tried to meet those challenges:

> We chose to focus first on breaking the [student's] habit of depending on patterns. Not only did these students become aware of their own abilities as they became involved in figuring out what to do when they worked with these irregular number systems, but their teachers in other courses, whom we asked to tackle the same problems the students were successfully handling, discovered, some with near disbelief, that the students were far more capable than had previously been apparent. When this happens the notion that higher expectations will produce the sought-after improvement in performance becomes more than just a good idea: teachers really do expect more of these students because they really believe the students can do more.

I cannot express the distinction between a challenge and an excuse any more eloquently than Orr. We need to draw that distinction for ourselves, our colleagues, and the administrators with whom we work. Though there are no magic cures to the problems that beset the mathematics education of minority students, we are all in this together.

It takes time to identify the challenges we face; it also will take time to find solutions to those challenges. We must tap the creative resources of teachers. Equally important, we need visionary leaders within our schools and professional organizations and researchers who can articulate the challenges facing us in the mathematics education of women and minority students. Our world is becoming more complex. Not only must the mathematics we teach meet the social and economic demands of that world, but also we must take steps to ensure that *everyone* participates in that world.

REFERENCES

Arias, M. Beatriz. "The Context of Education for Hispanic Students: An Overview." *American Journal of Education* 95 (1986): 26–57.

Center for Education Statistics. *The Condition of Education.* Washington, D.C.: U.S. Government Printing Office, 1987a.

———. *Digest of Education Statistics.* Washington, D.C.: U.S. Government Printing Office, 1987b.

Clark, Christopher M., and Penelope L. Peterson. "Teachers' Thought Processes." In *Handbook of Research on Teaching* (3d ed.), edited by Merlin C. Wittrock, pp. 255–96. New York: Macmillan Publishing Co., 1986.

Dossey, John A., Ina V. S. Mullis, Mary M. Lindquist, and Donald L. Chambers. *The Mathematics Report Card: Are We Measuring Up?* Trends and Achievement Based on the 1986 National Assessment. Report no. 17-M-01. Princeton, N.J.: Educational Testing Service, 1988.

Espinosa, Ruben, and Alberto M. Ochoa. "Concentration of California Hispanic Students in Schools with Low Achievement: A Research Note." *American Journal of Education* 95 (1986): 77–95.

Grubb, Norton. *Educational Reform and the New Orthodoxy: The Federal Role in Vitalizing the U.S. Labor Force.* Prepared for the Joint Economic Committee, U.S. Congress. 1986. [Photocopy available from author.]

Hispanic Policy Development Project. *Closing the Gap for Hispanic Youth: Public/Private Strategies.* Washington, D.C.: The Project, 1988.

Hodgkinson, Harold L. *All One System: Demographics of Education, Kindergarten through Graduate School.* Washington, D.C.: Institute for Educational Leadership, 1985.

Kennedy, Mary M., Richard K. Jung, and Martin E. Orland. *Poverty, Achievement, and the Distribution of Compensatory Education Services.* Interim Report from the National Assessment of Chapter 1, OERI. Washington, D.C.: U.S. Government Printing Office, 1986.

Johnston, William B., and Arnold E. Packer. *Workforce 2000: Work and Workers for the Twenty-first Century.* Indianapolis: Hudson Institute, 1987.

MacCorquodale, Patricia. "Mexican-American Women and Mathematics: Participation, Aspirations, and Achievement." In *Linguistic and Cultural Influences on Learning Mathematics,* edited by Rodney R. Cocking and Jose P. Mestre, pp. 137–60. Hillsdale, N.J.: Lawrence Erlbaum Associates, 1988.

Myers, David E., and Ann M. Milne. "Effects of Home Language and Primary Language on Mathematics Achievement." In *Linguistic and Cultural Influences on Learning Mathematics*, edited by Rodney R. Cocking and Jose P. Mestre, pp. 259–93. Hillsdale, N.J.: Lawrence Erlbaum Associates, 1988.

National Alliance of Business. *Employment Policies: Looking Ahead to the Year 2000*. Washington, D.C.: The Alliance, 1986a.

________. *Youth 2000: A Call to Action*. Report on a national leadership meeting held 10 June 1986. Washington, D.C.: The Alliance, 1986b.

National Commission on Excellence in Education. *A Nation at Risk: The Imperative for Educational Reform*. Washington, D.C.: U.S. Government Printing Office, 1983.

National Council of Teachers of Mathematics. *Curriculum and Evaluation Standards for School Mathematics*. Reston, Va.: The Council, 1989.

Oakes, Jeannie. *Opportunities, Achievement and Choice: Issues in the Participation of Women, Minorities and the Disabled in Science*. Santa Monica, Calif.: Rand Corp., 1987.

O'Malley, J. Michael. *Children's English Services Study: Language Minority Children with Limited English Proficiency in the United States*. Rosslyn, Va.: National Clearinghouse for Bilingual Education, InterAmerica Research Associates, 1981.

Orr, Eleanor, W. *Twice as Less: Black English and the Performance of Black Students in Mathematics and Science*. New York: W. W. Norton, 1987a.

________. "Does Black English Hinder Learning Mathematics?" *Washington Post*, 1 November 1987b.

________. "Commentary: Black English as an Obstacle to Learning." *Education Week*, 2 December 1987c.

Oxford-Carpenter, Rebecca, Louis Pol, David Lopez, Paul Stupp, Murray Gendel, and Samuel S. Peng. *Demographic Projections of Non-English-Background and Limited-English-Proficient Persons in the United States to the Year 2000 by State, Age, and Language Group*. Rosslyn, Va.: National Clearinghouse for Bilingual Education, InterAmerica Research Associates, 1984.

Resegregation of Public Schools: The Third Generation, a Report on the Condition of Desegregation in America's Public Schools. Network of Regional Desegregation Assistance Centers. Portland, Oreg.: Northwest Regional Educational Laboratory, 1989.

Romberg, Thomas A., and Thomas P. Carpenter. "Research on Teaching and Learning Mathematics: Two Disciplines of Scientific Inquiry." In *Handbook of Research on Teaching* (3d ed.), edited by Merlin C. Wittrock, pp. 850–73. New York: Macmillan Publishing Co., 1986.

Strike, Kenneth A. "Is There a Conflict between Excellence and Equity?" *Educational Evaluation and Policy Analysis* 7 (1985): 409–16.

Tikunoff, William J. *Applying Significant Bilingual Instructional Features in the Classroom*. Rosslyn, Va.: National Clearinghouse for Bilingual Education, 1985.

U.S. Bureau of the Census. *Projections of the Hispanic Population: 1983 to 2080*. Current Population Reports, Population Estimates and Projections, Series P-25, No. 995. Washington, D.C.: Government Printing Office, 1986.

Veltman, Calvin. *The Future of the Spanish Language in the United States*. Washington, D.C.: Hispanic Policy Development Project, 1988.

17

Teaching Mathematics: A Feminist Perspective

Suzanne K. Damarin

> As a pluralistic, democratic society, we cannot continue to discourage women and minority students from the study of mathematics. . . . We challenge all to develop instructional activities and programs to address this issue directly.
>
> *Curriculum and Evaluation Standards for School Mathematics*

THE current wave of feminism, only two decades old, has sparked debate, discussion, and change in virtually every aspect of our society. Looking back over this period, we can see in our schools many changes brought on by this movement. Perhaps the most salient differences are in athletic departments and extracurricular activities. Within the academic areas, additions and modifications have been made to the subject matter, course of study, and lesson plans in many subjects, most notably in English and history. All our textbooks and teaching materials are now judged by both publishers and adoption committees to assure that they are free of "sexist bias." Within the area of mathematics, increased research on women and mathematics, including work on attitudes toward mathematics and the role of course taking in mathematical achievement, has led to some changes in the mathematical experiences of girls and young women. In recent years the scientific community has actively begun to recruit young women to the study of advanced science, including the mathematical sciences. Recalling this impressive list of changes, the reader might feel that the problems of equal treatment of female students and the issues raised by feminists have been solved for all time. A closer analysis, however, shows that although the changes made to date have often been controversial and even painful for some, they have also been the easiest and most obvious changes that could be made.

The purpose of this article is to address, from a feminist perspective, the need for further change in the teaching of mathematics. In order to do so, it is essential to discuss first some of the principles of feminist thought. A fundamental tenet of feminist thinking is that we must know, and come to grips with, our own biases and belief structures. We must think about how these affect our actions before we can make fundamental changes; discussion with others can help us to see these effects and the means of change. This idea was the underlying principle of the "consciousness raising" movement of the 1970s when women got together and through discussion elevated their subconscious thoughts and feelings to a conscious level in order to sort them out, to select those they chose to act on, and, finally, to change their behaviors and their lives. In this article several areas for "consciousness raising" among mathematics teachers will be identified and discussed.

Another point of importance to feminist thinking is the idea that our society—including all aspects of our schools, our curricula, our graded course(s) of study, many of our teaching methods, and even the subject matter of mathematics(!)—was created by *men* (primarily) and reflects the life experiences and goals of men. These institutions and practices began in a period when many people, including many scientists and philosophers, considered women to be incapable of rational thought. Although most people do not hold that belief today, vestiges of this history remain at the core of how we think and of much that we do.

To recognize that mathematics has masculine historical roots is not to suggest that women cannot or should not learn it today, nor that mathematics as we teach it is irrelevant to the lives of modern women. Rather, this recognition can help us to see the depth of the "gendering" of the subject and to identify ways in which we might look for gender issues in instruction. For example, it suggests that we examine mathematics instruction in light of the findings by psychologists and sociologists that boys are socialized to be more aggressive than their female peers (Linn and Petersen 1985).

When we seek reflections of aggressiveness, we find it abounding in instructional practice. Our vocabulary reflects goals of *mastery* and mathematical *power.* We teach students to *attack* problems and to apply *strategies.* Our instructional strategies include *drill* and the use of many forms of *competition.* We are advised to *torpedo* misconceptions and to build concept *hierarchies.* In short, the ways we think about, talk about, and act out our roles as teachers of mathematics are heavily influenced by the masculine roots of the subject. It is true that some terms used in relation to mathematics are associated with the feminine; *beautiful* proofs, *elegant* proofs, and *nice* solutions are all part of mathematics, the *queen* of the sciences. These terms refer, however, to mathematics as already completed, not to mathematics as an activity to be engaged in. They do not suggest for girls an active role in the "learning by doing" of mathematics. The idea that mathematics is the *handmaiden* of the sciences (Bell 1937) is, indeed, a feminine reference to activity, but it suggests a low and demeaning level of activity, clearly not one to which young women might aspire.

"You can't change history," the reader might say, and this feminist would surely agree. But we can use the "raised consciousness" of this linguistic history to help us seek ways to meet the challenge of the NCTM's *Curriculum and Evaluation Standards* (National Council of Teachers of Mathematics 1989) to change current practice. As a first step we might think about, and discuss, what it would mean if instead of working toward students' mastery of facts and concepts, we worked toward students' *internalization* of them. Instead of leading students to attack problems, why not *interact with* them? We might *share* problems and work *cooperatively* toward their solution. Rather than torpedoing misconceptions, we might share the lack of fit between two ideas, making of it a problem for cooperative *resolution.* Through these procedures our students might build, not hierarchies, but *networks* of concepts, facts, problems, and procedures.

With the modification of the language of instruction and the use of a more cooperative approach to realistic problems can come a decrease in the psychological distance of many students, especially girls and women, from the subject matter. Teachers will also find that the task of de-emphasizing competition in instructional practice is one that may never be finished. New words will appear as problematic; new insights will reveal the sources of distance between some students and mathematics. Many questions will occur; everybody's list of questions will be different, but here are some examples:

- How deeply rooted in competition is mathematics instruction? Students compete with each other, with themselves (what does that mean?), with the computer, with the clock. . . . With what else?
- How can students be encouraged to use "sharing time" to share their own mathematical problems of the economics of allowances, the allotment of time to activities, or quantitative goal setting?
- What are some good problems involving quadratic equations which do not involve trajectories (which I tend to associate with bullets and bombs)?

Although not all these questions will be resolved quickly, talking about them with colleagues and friends keeps them elevated in consciousness. It is hard to go from a good discussion of issues such as these to the day-to-day routines of rule-and-example teaching or drill and practice. Consciousness raising does change one's life!

SEX DIFFERENCES AND MATHEMATICAL ABILITY

Issues that separate girls and women from mathematics and mathematical sciences are not all rooted in the language and methods of instruction. As noted above, a century ago many influential writers believed that women were incapable of mathematics. Even into this century this belief was so strongly held within the community of mathematicians that when Emmy Noether (1882–1935) proved herself to be a mathematical genius, she was given the (German) masculine title *Der Noether;* it was easier for her colleagues to deny that she was of the female sex than to give up what we now see as an extreme prejudice. Over the last century this extreme prejudice has been dislodged by the demonstrated ability of many women to learn and to teach mathematics and by the scientific study of psychology (Perl 1978).

The recent history of research on sex differences in mathematical ability and achievement is complex. The general trend of research findings provides increasing evidence that sex differences in overall mathematical ability are negligible or nonexistent (Deaux 1985; Hyde and Linn 1986). This trend is consistent with what many teachers, together with many critics of earlier research, believed (or wanted to believe) all along. However, findings that there are no sex differences in overall mathematical ability has neither ended the study of the question nor been widely publicized.

The public perception of scientific "facts" is determined, not by the whole body of scientific discovery, but by that portion that is highly publicized (Nelkin 1987). Therefore, the high level of publicity given to earlier studies (Benbow and Stanley 1980, 1983) as well as more recent studies (e.g., Blakeslee 1988) that purported to show strong sex differences are more influential on females' perceptions of their own mathematical ability than the unpublicized findings showing equality. In short, a part of the mathematical reality of the female student is the messages that she and others have received concerning her mathematical ability. The two burdens of these messages are (1) that the question of whether females are innately inferior in mathematics is a legitimate question meriting continued scientific study, and (2) that the answer has generally been reported to be yes. Even if the second message is being falsified by research, in the absence of publicity for findings of no difference, the power of this message remains intact unless we, as teachers, confront it.

One method of confronting this message with students is to bring research reports or newspaper and magazine articles on sex differences (and the letters to the editor that dispute them) into the classroom. Raise questions concerning them: Are the data well grounded and relevant to the question? What computations are involved? Do the numbers reported support the claims of the headline? Such discussions can help students both to "problematize" the question of sex differences in mathematical ability and to see that the application of mathematics to important real-world questions is not as cut and dried as their textbook examples.

Many in the field of mathematics education have been concerned with the cultural and psychological variables that are related to females' (and males') performance in mathematics. Numerous psychological variables and hypotheses have been studied in this regard. Studies have revealed relationships between females' performance in mathematics and their scores on measures of confidence, anxiety, fear of success, risk taking, attribution of success and failure to internal versus external factors, their perception of mathematics as a male domain, parental support for education, and related variables (see Fennema and Sherman 1977; Reyes 1984; Stage et al. 1985). However, the relatively small effects reported in many of these studies of individual variables have deflected attention from the *cumulative effects* of all these variables operating together and in concert with other conditions that affect females' learning.

Several new feminist psychologies of women address the cumulative effects of female socialization and seek a more holistic view of female thought. Belenky and colleagues (1986), for example, have identified stages of women's cognition. In this analysis women learn abstractions (such as mathematical principles) best if statements of rules are preceded by quiet observation, by listening to others, and by personal experiences that women can relate to the abstractions. The personal mathematical experiences through which females understand abstractions often differ from those of males. In Turkle's (1984) study, for example, both boys and girls programmed in Logo, but girls' programs were more personalized, less structured, and less "clean"; they were therefore less likely to be perceived by teachers as excellent. Despite their departure from the objectives of efficiency, nonredundancy, and "top down" problem solving, however, these programs gave girls their first real opportunity to engage the power of mathematics to solve personally defined problems.

Both the new NCTM standards and the increasing availability of the computer invite us to use fewer didactic lessons and to engage in more problem solving and hypothesis testing as classroom activities. As we make this shift in instruction, we need to consider both that numerous studies have shown that males outperform females on tests of problem solving and that we do not know why this is so. It is not clear that increased use of current approaches to problem-solving instruction will help females become better problem solvers. An equally plausible hypothesis is that the effects of all the variables cited above accumulate in problem-solving activities that require students to risk being "wrong" in their hypotheses, to rely on internal intuitions, and to cope with the possibility of success in a "male domain." As we move toward increasing the role of problem solving in the classroom, we need to explore both how we can help females deal with these phenomena and how we can provide opportunities that promote learning in ways consistent with the findings of Belenky and her colleagues.

All of this suggests that female students need a special kind of attention from teachers. Yet, findings of research into gender differences indicate that, on the average, teachers interact with individual female students less often than they interact with individual males. Moreover, their interactions with males are more frequently task relevant than their interactions with females (Fennema and Peterson, 1987). Thus, the students who are likely to need more interaction with their teachers are likely to receive less.

What Can Teachers Do?

If this situation is to be remedied, teachers must recognize it not only as a general problem but as a personal problem for individual attention. By making a quick tally at the end of each mathematics class for a few weeks, teachers can determine how frequently they interacted with students of each gender (as well as students of various racial or ethnic backgrounds). By making a list each day of a few students to talk with about mathematics, teachers can begin working toward balancing their interactions with students of both genders and of all groups. By working toward quantitative equality in the number of interactions with female and male students, teachers will have *begun* to make a difference. But a more fundamental issue is the content and quality of these interactions, which is a more difficult problem.

In addressing this problem, we can call on both colleagues and technology for help. Viewing the appropriate videotapes from the Multiplying Options and Subtracting Bias series (Fennema 1980) with colleagues, students, and parents can help alert everyone to some general issues. Making audiotapes and videotapes within the classroom and analyzing them alone or with colleagues can alert teachers to how they subconsciously replicate messages of bias. Once these subconscious activities are identified, conscious efforts to change them can be made. For example, if a teacher finds (as I did) that she allows male students more thinking time than females when responding to a question or problem, she can resolve to give girls more time—and devise a strategy for implementing this resolution. (What I did was to mentally recite a brief poem after posing a question and before moving on; after a while I became "naturally" less impatient and no longer needed the poem.)

An important contribution to the quality of teacher interaction with each female student is recognition of the fact and the importance of the affective backgrounds and conditions that each student brings to the study of mathematics. *To fail to recognize a student's anxiety, uncertainty, or concern about whether women are mathematically inferior is to deny an important part of the mathematical reality of the student.* Teachers must open discussions with young women concerning these issues. Beginning such discussions is tricky and requires considerable planning and thought. These discussions need room in our lesson plans, and some planning on how they will be started and continued. More than that, they require trust of a different quality than is needed for most mathematics instruction. A teacher must be willing to share personal beliefs and feelings and the findings from personal reflection and discussion. "You know, a lot of people have math anxiety. . . . Do you think everyone does? . . . I remember the time that. . . . Of course, once I learned it, I wasn't as anxious. . . . What do you think we could do about math anxiety in our class? . . . Can you believe that scientists used to think women couldn't do math? . . . Of course they never let them try. . . . My grandmother still believes it. . . ." The discussions will be different in every class.

Although there is no research evidence that such discussions will solve the problems of women and mathematics, there *is* ample evidence that through consciousness raising, group discussions, and even group therapy people have begun to resolve similar problems. Do not expect to see changes overnight! The goal of these discussions is *not* to create improvement on the next test score but rather to help young women (and all students) to see more clearly, and perhaps to change, their own approaches to mathematics. This goal, if met, will facilitate their full participation in a society that is increasingly mathematical.

As we enter the 1990s, we can anticipate that mathematics will find increasing application in all fields. Therefore, a major objective of mathematics instruction must be that all students *learn that they can learn* mathematics. Too often we give students the opposite message; we stress the importance of learning a particular topic at a particular time and hold out little hope for learning it later. Women (and minority students) especially are frequently advised to seek interests that do not involve mathematics. According to sociologist David Maines (see Fennema 1985), these messages can be especially debilitating to adolescent women who are struggling with the need to prepare for multiple roles in life (career, wife, mother) and often cannot devote to mathematics the single-minded attention that understanding and learning it requires.

Obviously we must stop discouraging young women, both overtly and covertly, from pursuing interests that involve mathematics. In addition, we must help them to see potential relationships of mathematics to their lives. Fortunately, there are increasingly many resources available to teachers to use in helping young women to expand their mathematical horizons (see Jacobs 1988). However, these resources must be supplemented by the affirmation of the mathematical potential of each girl and young woman. In summary, affirming that potential requires that we, as teachers, raise our own consciousness of the mathematical realities of our female students and work to change those realities.

REFERENCES

Belenky, Mary F., Blythe M. Clinchy, Nancy R. Goldberger, and Jill M. Tarule. *Women's Ways of Knowing*. New York: Basic Books, 1986.

Bell, Eric T. *The Handmaiden of the Sciences*. Baltimore: Williams & Wilkins Co., 1937.

Benbow, Camilla P., and Julian C. Stanley. "Sex Differences and Mathematical Ability: Fact or Artifact?" *Science* 210 (1980): 1262–64.

________. "Sex Differences in Mathematical Reasoning: More Facts." *Science* 222 (1983): 1029–31.

Blakeslee, Sandra. "Female Sex Hormone Is Tied to Ability to Perform Tasks." *New York Times*, 18 November 1988, pp. 1, 6.

Deaux, Kay. "Sex and Gender." *Annual Review of Psychology* 36 (1985): 49–81.

Fennema, Elizabeth. Multiplying Options and Subtracting Bias. 4 videotapes with guide. Reston, Va.: National Council of Teachers of Mathematics, 1980. (Also available by rental from Women in Mathematics Education, c/o SummerMath, Mount Holyoke College, 302 Shattuck Hall, South Hadley, MA 01075.)

Fennema, Elizabeth, ed. "Explaining Sex-related Differences in Mathematics: Theoretical Models." *Educational Studies in Mathematics* 16 (1985): 303–20.

Fennema, Elizabeth, and Penelope L. Peterson. "Effective Teaching for Girls and Boys." In *Talks to Teachers*, edited by David C. Berliner and Barak V. Rosenshine, pp. 111–25. New York: Random House, 1987.

Fennema, Elizabeth E., and Julia Sherman. "Sex-related Differences in Mathematics Achievement, Spatial Visualization, and Affective Factors." *American Educational Research Journal* 14 (1977): 51–71.

Hyde, Janet S., and Marcia C. Linn. *The Psychology of Gender: Advances through Meta-Analysis*. Baltimore: Johns Hopkins University Press, 1986.

Jacobs, Judith E., ed. *WME Bibliography of Mathematics Equity and Computer Equity*. South Hadley, Mass.: Women in Mathematics Education, 1988.

Linn, Marcia C., and Anne C. Petersen. "Facts and Assumptions about the Nature of Sex Differences." In *Handbook for Achieving Sex Equity through Education*, edited by Susan S. Klein, pp. 53–77. Baltimore: Johns Hopkins University Press, 1985.

National Council of Teachers of Mathematics. *Curriculum and Evaluation Standards for School Mathematics*. Reston, Va.: The Council, 1989.

Nelkin, Dorothy. *Selling Science: How the Press Covers Science and Technology*. San Francisco: W. H. Freeman & Co., 1987.

Perl, Teri C. *Math Equals: Biographies of Women Mathematicians and Related Activities*. Reading, Mass.: Addison-Wesley Publishing Co., 1978.

Reyes, Laurie Hart. "Affective Variables and Mathematics Education." *Elementary School Journal* 84 (1984): 558–81.

Stage, Elizabeth K., Nancy Kreinberg, Jacqueline Eccles (Parsons), and Joanne Rossi Becker. "Increasing the Participation and Achievement of Girls and Women in Mathematics, Science, and Engineering." In *Handbook for Achieving Sex Equity through Education*, edited by Susan S. Klein, pp. 237–68. Baltimore: Johns Hopkins University Press, 1985.

Turkle, Sherry. *The Second Self: Computers and the Human Spirit*. New York: Simon & Schuster, 1984.

18

African-American Students and the Promise of the *Curriculum and Evaluation Standards*

Lee V. Stiff

"Come to me," she called out. "Come over here."
And the young ones answered, "It's too close to the edge!"
"Come to me. I will protect you."
But again, the little ones were afraid and cried, "We might fall!"
"Don't worry," was her reply. "Come to me. I'll take care of you."
So they came near. She smiled, and then she pushed them! And as she knew
they would, the young eagles soared.

THE decade of the eighties was filled with discouraging reports of poor mathematics achievement among U.S. students. Among African-American students, depressed mathematics achievement was particularly pronounced. The evidence from NAEP (see Dossey et al. 1988) over the past four assessments demonstrates that African-American students consistently achieved about one proficiency level below their white counterparts. The most striking observation was that as age and level of proficiency increased, so did performance gaps between blacks and whites. Data from the 1986 assessment (table 18.1) show the level of mathematics proficiency of black and white students. These data clearly indicate that African-American students are not as successful as their white counterparts in school mathematics.

Unfortunately, the mathematics classroom is one of the most segregated places in American society. General mathematics classrooms contain disproportionate numbers of black students, whereas algebra II, geometry, and advanced mathematics classes mainly serve white students (Dossey et al. 1988; Matthews, Carpenter, Lindquist, and Silver 1984). Many mathematics

TABLE 18.1
Percentage of Students at or above the Five Mathematics Proficiency Levels: 1986

Levels of Proficiency	Age 9	Age 13	Age 17
	Black students		
Simple arithmetic facts	93.0	100.0	100.0
Beginning skills and understanding	53.3	95.5	100.0
Basic operations and beginning problem solving	5.4	49.4	86.0
Moderately complex procedures and reasoning	0.0	4.0	21.7
Multistep problem solving and algebra	0.0	0.1	0.3
	White students		
Simple arithmetic facts	98.9	100.0	100.0
Beginning skills and understanding	79.2	99.2	99.9
Basic operations and beginning problem solving	24.5	78.7	98.3
Moderately complex procedures and reasoning	0.7	18.6	58.0
Multistep problem solving and algebra	0.0	0.5	7.6

educators across the nation observe that African-American students now have less access to college preparatory courses in mathematics than before the mid-1960s. To use a familiar example, a significantly smaller percentage of black students took algebra I, geometry, and algebra II in North Carolina in 1987–88 than in 1963–64 (North Carolina Board of Education 1988; R. R. Jones, personal communication, 1989).

Many of the factors that affect the performance of black students, such as course enrollment patterns, role models, and significant others (e.g., counselors), have been shown to be important elements in dealing with the problem of poor mathematics achievement (Dossey et al. 1988; Matthews 1984; Welch, Anderson, and Harris 1982). It should not be surprising that the more mathematics courses students take, the better their performance on achievement tests will be; that the presence of African-American mathematics teachers in classrooms helps to formulate students' ideas about who excels in mathematics; that counselors who will not accept the practice of tracking blacks into vocational and remedial mathematics programs send the message that all children are expected to achieve. All too often, however, educators have relegated such "remedies" to subsets of black students (often identified as "talented") who must participate in special intervention programs existing outside of normal school experiences.

In general, little or nothing about school mathematics has been eliminated or modified to alleviate the poor performance by African-American students. One might conclude from this that the "problem" does not lie in the curriculum, existing teaching practices, staffing patterns, or the way we evaluate performance in mathematics. Rather, African-American students alone are responsible for their inability to achieve in the mathematics classroom. This, however, is in sharp contrast to the "no blame" perspective of *The Underachieving Curriculum* (McKnight et al. 1987), in which the mathematics curriculum and teaching practices are seen as possible causes of students' poor performance in mathematics achievement.

CULTURE AND SCHOOLING

There is a discrepancy between how two similar problems of underachievement by U.S. students in general and African-American students in particular have been addressed or left unresolved. How the problem of underachievement and underrepresentation in mathematics of any group of students is stated has a lot to do with the solution that will be fashioned. "Culturally diverse groups" have been described as "culturally deprived," "culturally disadvantaged," "marginal students," and the latest, "at-risk students." All these terms suggest that the students, their families, their homes, their communities, and so on, are somehow inadequate. This perception of inadequacy undoubtedly affects many teachers' perception of students' *cognitive abilities* (see Brophy and Good 1974; Neisser 1986). That much of Japanese society and culture is different from our own is seldom questioned. Yet, not once in the debate about underachieving students in the United States was it seriously suggested that the American culture is deprived or disadvantaged or that its students are cognitively impaired!

Important knowledge and skills that preschoolers bring to mathematics classrooms include ordering numbers, conservation of numbers, addition, counting, and reading and writing numerals. Black students are not likely to succeed in learning more formal mathematics without these and similar skills. Research has demonstrated that African-American children bring to the formal classroom setting the same basic intellectual competencies in mathematical thought and cognitive processes as their white counterparts (Ginsburg 1986). By the time black students enter kindergarten, virtually all of them are prepared to succeed in the school mathematics they will face. Since the difficulty that black students encounter in learning mathematics does not reveal itself before formal learning begins, it may be attributable to the experiences that blacks undergo in the school setting. Perhaps, *school mathematics* becomes the obstacle to success in the mathematics education process.

Schools must accommodate African-American children. School mathematics is more than the mathematical facts, concepts, and principles that students are asked to learn. From a sociocultural perspective, it is the expression of how European cultures view the world. The so-called analytical teaching and learning styles (Gilbert and Gay 1985) that are used in most of the classrooms in America reflect a preference for analytical thinking and systematic approaches to problems. Analytical teachers and learners try to place order and structure on the world to understand it. The manner in which mathematical concepts and principles are communicated reflects the importance of order to the mathematical system itself. The importance of directness, precision, and conciseness leads to a highly symbolic mathematical system that uses elaborate syntactical codes to represent the world. Individual achievement is prized, and therefore competition is acceptable and desirable behavior. It is as if the ecology of the mathematics classroom is a struggle among students for limited supplies of knowledge, skills, and understanding. Consequently, school mathematics, as currently practiced, sorts students throughout the education process.

African, Hispanic, and Asian cultural groups have organized differently, and attributed different importance to, the mathematical systems they use to explain the world (Ginsburg 1978; Hale 1982; Zaslavsky 1973). Although not identical, these mathematical systems address the functions of counting and the usefulness of manipulating concrete objects. They incorporate verbal and motor skills to communicate mathematical ideas and relationships. These so-called relational teachers and learners view the world as a unified environment with inherent order for which relationships (including interpersonal ones) are the focus (Gilbert and Gay 1985). African-Americans are more likely to approach problems holistically, defer (not neglect or omit) analytical investigations, and limit elaborate syntactical representations of the world. Relational teachers and learners prefer to use descriptive modes and tend to view the world in relative terms (in context). Benefit to the group is valued, and consequently cooperative learning is acceptable and desirable behavior. The ecology of the mathematics classroom in which cooperative learning is valued has unlimited amounts of knowledge, skills, and understanding to share, and every student is expected to partake of them.

The productivity of any teaching episode involving African-American students will be greatly diminished if teaching and learning styles that are compatible with black cultural norms are never employed. Poor performance by blacks in mathematics, then, may be the direct result of inappropriate teaching perspectives. Dossey et al. (1988) pointed out that mathematics instruction in 1986 continued to be the lecture style of presentations in which teachers relied heavily on the chalkboard, textbooks, and worksheets. In their words, "More innovative forms of instruction—such as those involving small group activities, laboratory work, and special projects—remain disappointingly rare." (p. 10) Such innovative forms of instruction would benefit African-American students.

TEACHING AND LEARNING MATHEMATICS IN THE 1990S

Mathematics classrooms operate under a set of values, orientations, and expectations. Many examples illustrate how black expression in mathematics classrooms produces negative feedback from teachers and frequently students (see Stiff and Harvey 1988). Consider the students who constantly move their seats so they can do their in-class mathematics assignments together, the consumer mathematics student who finds it necessary to talk about the new car he intends to buy before he can solve the car repair/maintenance problem, the algebra I student who wants the teacher to "explain this stuff" and is resistant to getting the information from reading the text, the tenth grader who presents her geometry proof using imprecise terms, implicit logical steps, and unnecessary comments about how clever she is, and the student who insists on going to the chalkboard to demonstrate that the solution to a system of equations is $(5, -3)$.

Each example represents behavior that may be acceptable at times, but many teachers who are continually faced with this type of behavior by black students often send the message that such conduct will not be tolerated. Students get the message: You are not the type of mathematics student we want. After all, the attributes of a successful student in school mathematics include working independently (not in support groups), being direct and concise (not telling tangential stories that may or may not be related to the problem), valuing direct and efficient methods of obtaining information (not the personal relationship that can be nurtured), using accepted (elaborate) syntactical discourse (not a conversational style), and responding in an orderly and structured way to classroom situations (not leaving one's seat to answer a question).

A significant effort is needed to modify mathematics curricula, existing teaching practices, staffing patterns, how students are advised and tracked, and the way performance in mathematics is evaluated if African-American students are to be successful in school mathematics. Curricula changes suggested by the *Curriculum and Evaluation Standards for School Mathematics* (National Council of Teachers of Mathematics 1989) call attention to these needs. The emphasis on making connections, mathematical communication, and cooperative learning, for example, should permit teachers to give a greater sense of security to all students but will make it equally important that teachers understand and value the elements of the culture that make African-Americans unique from others (see Havighurst 1978; Matthews 1984; Stiff and Harvey 1988). Similarly, the goals of the *Curriculum and Evaluation Standards* and the desire to increase among black students a sense of belonging point to the need for a racially and ethnically integrated teaching faculty (see Fiske 1986; Mercer 1982).

At the heart of the matter is whether the United States believes that the vast majority of all its students can reach high levels of mathematics achievement. The *Standards* document has staked out the position that all students can achieve and that all students should learn a certain core of mathematical knowledge. In the decade ahead, it is imperative that African-American students not be seen as an exception to the rule. Black students must be expected to take, and achieve in, the same courses as their white counterparts. This must be the goal of school mathematics.

The response to poor performance in mathematics, independent of the racial and ethnic background of students, should not be the removal of the opportunity to learn important and useful mathematics. The response should be to do whatever is necessary to reach the goals that have been set. Concerned mathematics teachers rarely accept failure from their advanced mathematics students. They usually assume that something more can be done. It may mean using different types of instructional activities acquired from other teachers, workshops, textbooks, and so on. With African-American students, teachers must be willing to apply the same set of heuristics!

Beyond what NCTM's *Standards* offers, when schools accept the responsibility for students who do not succeed and try to meet the needs of culturally diverse students in a manner consistent with the way such students live and learn, when teachers understand and care about their students and believe in them, failure is not acceptable.

REFERENCES

Brophy, Jere E., and Thomas L. Good. *Teacher-Student Relationships: Causes and Consequences*. New York: Holt, Rinehart, and Winston, 1974.

Dossey, John A., Ina V.S. Mullis, Mary M. Lindquist, and Donald L. Chambers. *The Mathematics Report Card: Are We Measuring Up?* Princeton, N.J.: Educational Testing Service, 1988.

Fiske, Edward B. "Minority Teaching Force Dwindling; Impact on Students Worries Educators." *The News and Observer* (Raleigh, N.C.), 9 February 1986, pp. 1A, 11A.

Gilbert, Shirl E. II, and Geneva Gay. "Improving the Success in School of Poor Black Children." *Phi Delta Kappan* 67 (1985): 133–37.

Ginsburg, Herbert P. "The Myth of the Deprived Child: New Thoughts on Poor Children." In *The School Achievement of Minority Children: New Perspectives*, edited by U. Neisser, pp. 169–89.

________. "Poor Children, African Mathematics, and the Problem of Schooling." *Educational Research Quarterly* 2 (1978): 26–44.

Hale, Janice. *Black Children, Their Roots, Culture, and Learning Styles*. Provo, Utah: Brigham Young University Press, 1982.

Havighurst, Robert J. "Structural Aspects of Education and Cultural Pluralism." *Educational Research Quarterly* 2 (1978): 5–19.

McKnight, Curtis C., F. Joe Crosswhite, John A. Dossey, Edward E. Kifer, Jane O. Swafford, Kenneth J. Travers, and Thomas J. Cooney. *The Underachieving Curriculum: Assessing U.S. School Mathematics from an International Perspective*. Champaign, Ill.: Stipes Publishing Co., 1987.

Matthews, Westina. "Influences on the Learning and Participation of Minorities in Mathematics." *Journal for Research in Mathematics Education* 15 (March 1984): 84–95.

Matthews, Westina, Thomas P. Carpenter, Mary M. Lindquist, and Edward A. Silver. "The Third National Assessment: Minorities and Mathematics." *Journal for Research in Mathematics Education* 15 (March 1984): 165–71.

Mercer, Walter A. (1982). "Future Florida Black Teachers: A Vanishing Breed." *Negro Educational Review* 33 (1982): 135–39.

National Council of Teachers of Mathematics. *Curriculum and Evaluation Standards for School Mathematics*. Reston, Va.: The Council, 1989.

Neisser, Ulric, ed. *The School Achievement of Minority Children: New Perspectives*. Hillsdale, N.J.: Lawrence Erlbaum Associates, 1986.

North Carolina Board of Education. *Statistical Profile, North Carolina Public Schools*. 14th ed. Raleigh, N.C.: The Board, 1988.

Stiff, Lee V., and William B. Harvey. "On the Education of Black Children in Mathematics." *Journal of Black Studies* 19 (1988): 190–203.

Welch, Wayne W., Ronald E. Anderson, and Linda J. Harris. "The Effects of Schooling on Mathematics Achievement." *American Educational Research Journal* 19 (1982): 145–53.

Zaslavsky, Claudia. *Africa Counts: Number and Patterns in African Culture*. Boston: Prindle, Weber & Schmidt, 1973.

19

Increasing the Achievement and Participation of Language Minority Students in Mathematics Education

Gilbert Cuevas

UNDERSTANDING and mastering English as the language of instruction appears to be a unique factor that language minority students need to overcome barriers that contribute to underachievement and underrepresentation in mathematics education (Cuevas 1984; NCTM 1987). Although some of these students enter school quite functional in English, many do not. In view of the linguistic needs of language minority students, this article will focus on issues and instructional suggestions that integrate the teaching and learning of language and mathematical content. (See also the article by Curcio in this volume.)

SOME ASSUMPTIONS AND GENERAL SUGGESTIONS

The integration of language skills and the teaching and learning of content involves understanding the nature of language and the role it plays in learning. The following three assumptions need to be made before addressing specific instructional strategies:

- Language must be discussed in terms of its component skills—listening comprehension, reading, writing, and speaking. Listening and reading are considered *receptive* language skills; writing and speaking are *productive*. Activities that address these areas need to be incorporated as part of the instructional processes. To illustrate, consider these observations of a sixth-grade teacher in a classroom with a large group of limited-English-proficient students. At the beginning of a unit on measurement, small groups of students measured objects using nonstandard units. Instructions were given in English, with the more proficient students assisting the less proficient. The teacher made a list of the vocabulary and expressions students needed to carry out subsequent activities (e.g., "Find the length of . . ."; "The [object] is . . . units long"). While the groups reported their findings verbally, the instructor wrote down the phrases the students used (with a bit of editing). These sentences were written on the board, read by the teacher, repeated orally by the students, and then copied by the class. In later sessions the teacher took time to carry out brief exercises in which the students wrote variations of these sentences. In each instance physical objects were used to reinforce the ideas contained in each of the sentences. The instructor was careful to integrate concept and language development through verbal, reading, and writing activities.

- Teachers and learners need to distinguish between the language used in daily communication and the "language of mathematics." Mathematical language is composed of a variety of linguistic structures (e.g., vocabulary) and functions (e.g., grammatical patterns, meanings). One of the characteristics of the language of mathematics is its lack of redundancy. Language used in everyday communication is redundant, that is, we do not need to know the meaning of every word or phrase used, since the message is repeated in the context of the communication. This is not true with mathematical language. Words and phrases have definite meanings and represent ideas that if not understood prevent the learner from fully understanding the message conveyed.

- The student's native language plays a role in the development of mathematical skills and concepts. From an intuitive point of view, instruction carried out in a language the student fully understands facilitates learning. From a research point of view, the relationship between language proficiency (in this context the student's first language) and content learning is a complicated one. Research results appear to support the notion that concepts and skills can be reinforced when language minority students have an opportunity to discuss them in their native language (Saville-Troike 1984; Cummins 1984; Hakuta 1986). Mathematics instruction needs to be structured so that language minority students have the choice of using their native language, if needed. This implies the use of collaborative learning groups, small-group discussions, and other strategies that emphasize communication among students and teachers. See NCTM's *Curriculum and Evaluation Standards* for elaborations of these ideas (National Council of Teachers of Mathematics 1989).

A SAMPLE OF INSTRUCTIONAL STRATEGIES

The instructional suggestions that follow focus on the development or reinforcement of language skills in the context of learning mathematics. The activities represent a sample of strategies the author has used, observed in classrooms, or found in the professional literature. The suggestions have been grouped into general instructional approaches and specific activities that are adaptable to different grade levels.

General Suggestions

Teachers often express concern over the obvious communication problems of limited-English-proficient students. Frequently these students are placed with teachers who are bilingual, but many times they are assigned to monolingual English teachers. The following suggestions address some general strategies for dealing with situations in which the native language of the students may or may not be understood by the teacher.

- "What if I do not speak the students' native language and they do not speak (or are not proficient in) English?" Do not panic! Here are some things you can do:

Be familiar with each student's educational background (Peck, Simmons, & Stark 1988). Find out what the native language of the students is, their country of origin (if immigrants), and a few facts about their culture or history. Many immigrant students come to school with varying degrees of proficiency in mathematics. For some of these students, the way in which they have learned mathematics is different from the manner taught in the United States. For example, Haitian students use the "additive approach" when subtracting. To compute 45 − 27, a Haitian student will say, "Seven plus what equals fifteen?"

There is a need to control the range of vocabulary and use of idiomatic expressions incorporated into the lessons (Jorgensen 1988). This can be achieved by using a "Teaching Script" exercise. Before class, write down in the form of a script (Teacher: . . .; Student: . . .) a *summary* of the types of interactions you want the class to carry out; what you will say to the students; the questions you will ask them; the information you will write on the board; and the written material you will give them. Also, write down the responses you wish the students to give. Once written, examine the script for words, phrases, and symbols you think will be troublesome to the students because of deficiencies in either their English language proficiency or their mathematical knowledge. Then either modify the language or incorporate some activities to teach the language. Teachers who have used the technique report that although it is quite a time-consuming task the first time it is done, the time required is reduced with practice.

- "What if I speak the students' native language?"

A knowledge of the students' language and cultural background is equally important in this situation.

Are the students literate in their native language? If so, materials written in the language may be given to them to carry out class activities, to introduce ideas, or to assess skills. Although a variety of mathematics materials is written in Spanish, some can be found in other languages as well. An oral approach with pictorial and manipulative materials can be used if the students are not literate in their native language. Coordination of activities with a bilingual class can also be used to introduce and develop literacy skills in mathematics in the student's native language.

Are there any cultural idiosyncrasies I need to be aware of? In general, this question concerns the assumptions we make about cultural background when addressing "Hispanic" or "Asian American" students. Many of the students in each of these groups have roots in countries with widely different cultures. It is ***not*** safe to assume familiarity with the culture and language of a particular group of students just because of the label we place on them. Get to know them—students are always willing to share information about their backgrounds. However, be sensitive to their privacy when asking questions.

Specific Suggestions

The following activities could be incorporated into daily classroom instruction:

Schedule part of each lesson for a review of the terms, words, and phrases used in previous lessons that are related to the lesson of the day (Jorgensen 1988). This will clarify or reinforce the langauge and concepts needed for the understanding of new material. To accomplish this, have students verbalize or write their understanding of terms or concepts. Examples include the following activities:

a) A matching exercise that stresses receptive language skills—in this example the context is measurement units (it is assumed the students have been exposed to concepts and terms such as standard and nonstandard units, weight, volume, etc.)

Directions: Match the item in Column A with the appropriate items in Columns B and C.

Column A	Column B	Column C
1) 21 steps	Standard Nonstandard	Weight Volume Distance Time
2) 200 pounds	Standard Nonstandard	Weight Volume Distance Time

3) 75 truckloads	Standard Nonstandard	Weight Volume Distance Time
4) 35 minutes	Standard Nonstandard	Weight Volume Distance Time

b) An activity that requires students to write or verbalize the meaning of certain terms (language production skills).

Directions: For each of the terms given below, write a sentence or two (or a short paragraph if needed) describing the term.

Term	**Written Description**
1. average	
2. computer program	
3. less than	
4. area	

A number of general strategies underlie these instructional illustrations:

- Whenever possible, use the students' past experiences with a concept or term to assist in the understanding of the idea (Garbe 1985).
- Have students share with one another their "definitions" or descriptions of terms before asking them to make presentations to the whole class.
- Use reality-based examples, which can facilitate the students' understanding and verbalization of the concepts and terms (Curcio 1985). If possible, use manipulatives. (This is a must in the primary grades!)

Give students opportunities to "talk about mathematics." NCTM's (1989) *Curriculum and Evaluation Standards for School Mathematics* emphasizes the need to "provide opportunities for [the students] to talk mathematics" (p. 26), and "to use language to communicate their mathematical ideas" (p. 78). This can be accomplished through the verbalization of processes used to solve problems. Students who may not be quite confident about their English language skills can be assisted by asking them questions to guide them. For example:

1. What are the important facts or conditions in the problem?
2. Do you need any information not given in the problem?
3. What question is asked in the problem?
4. Describe how you solved the problem.
5. Do you think you have the right answer? Why? Why not?
6. How did you feel while you were solving this problem?
7. How do you feel after having worked on the problem?

(Charles, Lester, and O'Daffer, 1987, pp. 24–25)

The primary purpose of the activity is to give students an opportunity to communicate orally the processes they have used to solve problems. The questions above may be asked individually or in small groups. This will alleviate students' language anxieties. Let them use their native language in these small groups if their English language skills do not allow them to fully communicate the process (appoint a group "translator" if you are not proficient in the students' native language). In addition, you may wish to make notes to be later addressed in class concerning points that were not expressed clearly.

Stress reading and writing skills in the mathematics lesson. Although this is an area that needs to be addressed with all students, language minority children—and especially those with limited English proficiency—particularly need reinforcement of reading and writing skills. To illustrate, an activity such as the following might be used:

Directions: Read each group of phrases or sentences. Then form a word problem by writing them in order. Use the blank lines on the right to write the problem.

1. for 6 hours on Saturday
   How much did she earn
   Helen worked
   at $4.75 per hour.

   ____
   ____
   ____
   ____

2. does he have?
   Francisco has some dimes
   How many coins of each kind
   and quarters
   totaling $4.05
   He has five more quarters
   than dimes.

   ____
   ____
   ____
   ____
   ____
   ____

(Crandall et al. 1986)

There is a need to provide language minority students with the skills to become proficient in the language used for mathematics instruction. The lack of appropriate language skills can be an obstacle to all students but particularly so for language minority students. It is by integrating the development of language in the context of mathematics instruction that we may give these students the "key that unlocks the door to the world of mathematics."

REFERENCES

Charles, Randall, Frank Lester, and Phares O'Daffer. *How to Evaluate Progress in Problem Solving.* Reston, Va.: National Council of Teachers of Mathematics, 1987.

Crandall, JoAnn, Teresa Corasiniti Dale, Nancy Rhodes, and George Spanos. *English Language Skills for Basic Algebra.* Washington, D.C.: Center for Applied Linguistics, 1986.

Cuevas, Gilbert J. "Mathematics Learning in English as a Second Language." *Journal for Research in Mathematics Education* 15 (March 1984): 134–44.

Cummins, James. "The Role of Primary Language Development in Promoting Educational Success for Language Minority Students." In *Schooling and Language Minority Students: A Theoretical Framework,* edited by D. P. Dolson, pp. 3–50. Los Angeles: Evaluation, Dissemination and Assessment Center, California State University, Los Angeles, 1984.

Curcio, Frances R. "Making the Language of Mathematics Meaningful." *Curriculum Review* 24 (March/April 1985): 57–60.

Garbe, Douglas G. "Mathematics Vocabulary and the Culturally Different Student." *Arithmetic Teacher* 33 (October 1985): 39–42.

Hakuta, Kenji. *Mirror of Language: The Debate on Bilingualism.* New York: Basic Books, 1986.

Jorgensen, Donna. "Language Development Increases Conceptual Understanding." Presentation made at the 29th Annual Meeting of the California Mathematics Council, November 1988, Long Beach, Calif.

National Council of Teachers of Mathematics. *Curriculum and Evaluation Standards for School Mathematics.* Reston, Va.: The Council, 1989.

________. "Mathematics for Language Minority Students." Position statement. *NCTM News Bulletin* 23 (May 1987): 9.

Peck, Sharon K., Pauline E. Simmons, and William Stark. *Math in a Limited English World.* Lansing, Mich.: The Lansing School District, 1988.

Saville-Troike, Muriel. "What Really Matters in Second Language Learning for Academic Achievement?" *TESOL Quarterly* 18 (1984): 199–219.

20

Cultural Power and the Defining of School Mathematics: A Case Study

Brian F. Donovan

IN CONSIDERING the influence of culture on the teaching and learning of mathematics, we should keep in mind that ultimately such influences are not about numbers and statistics but about those persons involved in the schooling process—in particular, students and teachers. The purpose of this article is to describe how cultural influences affect several aspects of the teaching and learning of school mathematics in a mathematics program in a "culturally deprived" suburban area of a large Australian city. The program was designed to enrich the mathematics of students at an elementary school declared by government authorities to be disadvantaged. The school is referred to here as Humble Street School. Culture was viewed as connected with the social relations of class, gender, and ethnicity. It was seen, in a study by Donovan (1983), as being lived out in the everyday social and political lives of students, parents, teachers, and administrators within a school that was itself situated in a wider community and world of unequal power relations. The mathematics program was defined not in words of a text but rather in a struggle of social groups.

HUMBLE STREET SCHOOL STUDY

Humble Street School became eligible for special grants because of its "deprived" status, and it was through these grants that the applied mathematics program, which the school labeled Enrichment Mathematics, was funded. These special grants were intended to increase educational opportunity in the most underprivileged neighborhoods, to make schooling more meaningful to students, and to encourage such schools to become more open and be better linked with their communities.

The descriptions that follow are based on classroom observations and interviews with the principal, six classroom teachers covering all grade levels, and the coordinator of the program. Interviews were conducted with the parents of five children: Australian and Greek parents, working-class and middle-class parents, and a single parent. The study also included interviews with significant others.

The applied mathematics program was developed by mathematics curriculum specialists, tested in selected schools, and then widely disseminated. The content consisted of activities in area, length, volume, mass, perimeter, geometry, time, money, and graphs. Separate texts for each of these topics were produced for teachers. The teaching of each topic was based on the idea that students learn by doing, and it emphasized activities for individuals and groups to do both in and outside the classroom. Manipulatives were to be used extensively. The program was open to organization by a single teacher within one or more classrooms or by several teachers across classes. It was assumed that teachers had control over the selection and organization of the program activities and that differential pacing for students would be encouraged.

The program was analyzed under three themes, each involving aspects of power relations: knowledge and cultures, instruction and cultures, and resistance and compliance.

KNOWLEDGE AND CULTURES

There were conflicting notions of achievement embodied in the words *enrichment mathematics* and *basics* which seemed to be linked to social class. An enriched curriculum suggests an addition of high value or quality, an in-depth treatment, and increased variety of experience. A basic curriculum, however, suggests the minimum requirements and fundamental experiences. Middle-class professional parents expressed the importance of enriched or extended experiences. One such parent asserted,

> I'm not so much worried about how they're taught, whether the computer way, or the long division way, or another form, as long as they understand there are other ways of it happening, and they don't get caught out in "there is the one and only way." Their minds then are not open to search out other ways in which we can work out maths.

In contrast to this, the working-class parents emphasized the importance of accepting authority, a moral code, and the learning of basic skills. One such parent said,

> If they're told to be quiet, I just hope my kids behave well. I think that goes a long way when they grow up as a citizen. Learn right from wrong. On the reports, to me, behavior goes a long way.

A parent of Greek origin commented,

> If you go in the rooms in Greece, everybody sit quietly. No talk, just listen to teacher. But me, I go sometimes into rooms here I see kids play here, play here, play here. If you are in grade 2 in Greece you understand everything in times

> table. How many twos in two. And my daughter now in grade 4 and she still no understand tables. Maybe my fault my kids, I don't know. Teacher ask here how many five in five. They count on fingers. Maybe that allowed in mathematics. . . . That's nice for them grade 1, grade 2, but not for grade 5, grade 6.

The mastering of basic number skills was a dominant concern for the working-class parents, whereas for the middle-class professional parents socializing experiences and "making learning exciting" were most important. Middle-class professionals tended to break down divisions between work and play in schooling so that the idea of enriched learning experiences was clearly an integral part of the intended curriculum. In-school and out-of-school knowledge were seen to significantly overlap. For the Australian working-class parents, school was a workplace and play was for outside of schooltime. For middle-class parents, homework was tolerable only if the schoolwork of a child was below standard. However, for the Greek working-class parents, homework not only assisted children to reach standards necessary for continued schooling but linked school to home by giving parents an opportunity to become familiar with the knowledge taught. Teachers were aware of the Greek parents' concern but generally did not respond to it or, when doing so, directed students to learn certain routines and facts that were likely to satisfy the parents. The views of teachers and middle-class professionals that there should be little homework was consistent with their notion that a major role of an elementary school was to provide opportunities for children to socialize.

The working-class parents did not value knowledge any less than the middle-class professionals. However, the particular knowledge that each social class held to be important and the work conditions considered suitable for its acquisition were different. It would be fallacious to ascribe certain views to a social class as though that class constituted some homogeneous grouping having the same orientation to knowledge and work. Although, for example, there were similarities in working-class views on homework, there were also differences. Working-class parents did have strong views on the appropriate knowledge and work for school mathematics, but their views did not have as much impact on the development of programs as middle-class professional views and action. However, although middle-class professionals dominated school curriculum committees and were able to gain funding for Enrichment Mathematics and other projects, this did not ensure that the programs were enriched in practice. Although there was some consonance between the views of school staff and middle-class professionals on elementary schooling as being primarily a socializing experience, there were contradictory elements in the meanings teachers gave to Enrichment Mathematics. While enrichment was being discussed, organizational arrangements were often a more pervasive force in shaping the mathematics program.

INSTRUCTION AND CULTURES

In the official language of the school, great emphasis was placed on catering to each child as an individual. Reference was made throughout the school policy to the "child" and "each child," giving a sense of the worth, uniqueness, and potential of students. "Catering for individual differences" always referred to academic outcomes.

The coordinator of the applied mathematics program saw her role as being a selector of activities, an organizer of appropriate materials, and a guide to students. In her interaction with students, she responded differently to boys and girls, the interested and uninterested, and the "brighter" and "slower" students. She was very conscious of spending too much time with boys, whom she felt were too demanding. On the notice board at the side of the room, she had pinned a newspaper clipping with the title "Boys Are Teachers' Pets." The article referred to research that stated that boys receive two-thirds of teachers' attention, taunt the girls without being punished, and receive praise for sloppy work that would not be tolerated from girls. In her own efforts to give greater time to interacting with girls, she frequently did not attend to boys' off-task and deviant behavior. The coordinator also described the division between boys and girls in cultural terms:

> The boys did not like working with girls. This . . . was partly related to boys' ages but also to their ethnic backgrounds. In Greek culture particularly, girls were looked down on by boys.

She allowed boys and girls to be seated separately so that she could better control her interaction time with girls and minimize the interference of disruptive behavior from some boys.

There was a strong relationship in teachers' perceptions between social class and the degree of compliance in task engagement. The following teachers' comments illustrate this relationship:

> If you take the more competent children in this grade who are generally the calmer, more sympathetic, more understanding children, who are both independent workers and can work quite well with a partner or in a group—all those children come from middle-class professional homes. (Grade P-1 teacher)

> I think [the kids who tend to say "I can't do that" come] from homes which are deprived in the way of caring. Nobody ever sits down and explains things to them. . . . Most of the high achievers are from professional families. (Coordinator of applied mathematics program)

> Just looking at them [working-class students in grades 2–3], they have concentration problems and emotional problems—they want more attention. They tend, when everyone else is working, to come up to you and want personal attention. (Grade 2–3 teacher)

The forms of differential instruction that were used often rendered the applied mathematics and the main mathematics programs irrelevant for many students, particularly working-class students. Although the principal and teachers were sympathetic toward these students, their "problems" were viewed as being so embedded in their social predicament that the most useful way to assist them was perceived to be relaxing the academic environment, which would alleviate stress and failure.

RESISTANCE AND COMPLIANCE

Teachers controlled students largely through work procedures but also through labeling and what Carlson (1982) has called "soldiering." At all grade levels there were students who were labeled "troublesome" and "problem children." This information was informally communicated between teachers, for example, at morning and lunch recesses. One strategy for controlling these students was to socially isolate them. When "a problem" developed, teachers would draw on the support of other students to focus on the need to make the "troublesome student" an outcast. The form of the isolation was being assigned to "the mat," "the corner," or to a room adjacent to the principal's office.

Carlson (1982) identified soldiering as one of the ways that teachers accommodate student resistance. It is the deliberate pacing of work at a level that is minimally taxing on students. At Humble Street School it could be seen in "individualized learning," where small numbers of students, grouped according to ability, were assigned a limited quantity of work that was usually accomplished well within the time given to the mathematics lesson. There was no coercion to have work completed, and mathematics activities were so fragmented that students just progressed to a new activity or unit in the succeeding lesson.

Two types of work stoppages were observed: one in which students quietly did no work over the whole or major part of a lesson, and the other, protest stoppages, where students demonstrated that their assigned work was too difficult. In most situations observed, a work stoppage was an individual or small-group phenomenon and involved "low ability" students. The quiet work stoppages were largely tolerated by teachers. In protest stoppages, teachers generally adopted one of two strategies: to ignore the situation or attempt to isolate a particular student. An instance of ignoring the situation was in a grade 3–4 activity involving manipulating triangular shapes:

> Shane threw his pieces down on the table and said he wouldn't do any more till he got help. He pushed his chair against the table, pushed the table in, and put his head down on the table; then he got up and left the group to play with blocks on another table. The teacher made no comment.

Sometimes, as in the following example with Tom, the teacher felt forced to intervene:

> Gayle [the teacher] helped Russell make a Christmas tree (from triangular shapes). But Tom called to her, "Miss, I can't do it." He had put two triangles together to form a square but had placed the others (six triangles) on top of each other so they appeared as one triangle atop the square. Gayle told Tom, "You can do it." Another child said, "He just wants to get out of work, Miss." Gayle went over to Tom, pushed some of his triangles nearly into the finished position but asked him to finish. He said, "I can't do it." Gayle said, "Just move these," as she moved several pieces together. Tom moved the two remaining triangles into place and called out to the other children, "Finished. I'm finished."

This points to the insight that students had of the "I can't do" behaviors. Although given as a laugh, one student's appraisal of the situation as a work stoppage seems to have been accurate. The teacher was already under pressure from the deviant behavior of others in the class. Tom exploited the situation, manipulating the teacher to achieve for himself the objective the teacher had set. The teacher recognized that she was being used but responded because of pressures other students were applying.

CONCLUSION

Culture is represented in the experiences, material artifacts, and practices of social groups. It is constructed in interactions of social groups with differential power. Working-class parents and Greek parents were silent in the face of the dominant taken-for-granted school arrangements and teacher ideology. "Culturally deprived" students participated in their own depression and to a large extent internalized inadequacies. Their power, evident in varied forms of resistance, was not enough to overcome their alienation and failure.

As teachers, we need to understand better the people and social groups who influence what happens in our classrooms, including the established structures and constraints that help determine access to, and success in, schooling. We also need to reflect on, and take more control over, everyday classroom and school practices that alienate and dehumanize students. The following suggestions are not meant to be exhaustive but are offered as starting points and general principles for working toward a more critical and empowering mathematics education:

- Begin with an understanding that students of all social classes, races, and both sexes have been and will be capable of mastering, developing, and using mathematics (cf. Gerdes 1985; Henderson 1981).
- Be aware that mathematics learning and teaching is packed with ideology: beliefs and practices that too often limit the creativity, critical expression, and human possibilities of students.
- Be more critical of mathematics texts, especially those developed for

mechanical teaching to "deposit" official mathematical words, procedures, and rules "into" learners.

- Understand and deal with "deficiency" and resistance in relation to their sociocultural context.
- Work to consider schooling a struggle for meaning, not just a job.
- Interact with students in a dialogue in which they are considered and treated as equals.
- Break down the cultural belief that there is only one "correct" answer to every question and that answer is given by authority in the form of teacher or text.
- Provide for collaborative group work and do not take for granted the belief that individual work and competitive relations are in students' best interests.
- Encourage a learning environment in which students become increasingly independent of the teacher, where they are respected and respect each other, and where they come to see the main objective in learning mathematics is to doubt, to inquire, to discover, to see alternatives, and most important of all, to construct new perspectives and convictions (Fasheh 1982, p. 3).
- Plan for and encourage students' construction of mathematics rather than teach it as an objective body of knowledge and skills to be learned, that is, see students as creators and recreators of mathematical ideas, not as consumers.
- Provide students with the opportunity to use their own reality as a basis for learning mathematics. This includes their language, for it is through language that they develop their own "voice," which is prerequisite to the growth of a positive sense of self-worth (cf. Freire and Macedo 1987).
- Be in contact with others interested in more critical education: teachers, parents, community people, and writers.

Teachers do not have exclusive power to define school mathematics. However, there are ways—as noted in this conclusion—in which they can challenge beliefs, practices, and taken-for-granted arrangements that perpetuate the disabling of students. There are many opportunities for struggling together with the "outsiders," "disadvantaged," and "culturally deprived" in creating a more empowering and critical education. Who defines mathematics and how it is defined will be problematic.

Teachers, though, can make a difference. Mathematics can be a means for understanding and transforming our world toward a more human place in which to live.

REFERENCES

Carlson, Dennis. " 'Updating' Individualism and the Work Ethic: Corporate Logic in the Classroom." *Curriculum Inquiry* 12 (1982): 125–60.

Donovan, Brian F. "Power and Curriculum Implementation: A Case Study of an Innovatory Mathematics Program." Ph.D. dissertation, University of Wisconsin–Madison, 1983.

Fasheh, Munir. "Mathematics, Culture and Authority." *For the Learning of Mathematics* 3 (2) (1982): 2–8.

Freire, Paolo, and Donaldo Macedo. *Literacy: Reading the Word and the World.* South Hadley, Mass.: Bergin & Garvey, 1987.

Gerdes, Paulus. "Conditions and Strategies for Emancipatory Mathematics Education in Undeveloped Countries." *For the Learning of Mathematics* 5(1) (1985): 15–20.

Henderson, David. "Three Papers." *For the Learning of Mathematics* 1(3) (1981): 12–18.

21

The Invisible Hand Operating in Mathematics Instruction: Students' Conceptions and Expectations

Raffaella Borasi

ALTHOUGH competence in mathematics is becoming a prerequisite for most careers, an increasing number of students seem unable to succeed in mathematics. *Mathematics anxiety* is widespread, and too many students avoid enrolling in mathematics courses unless they are strictly required. Although social and psychological factors are certainly at the root of this situation, the role played by students' views of school mathematics could also be crucial.

Consider, for example, the following image of mathematics, expressed by a bright math-avoidant woman (Buerk 1981) and supported by many of the high school students to whom it was shown (Borasi 1986b):

> Math does make me think of a stainless steel wall—hard, cold, smooth, offering no handhold, all it does is glint back at me. Edge up to it, put your nose against it, it doesn't give anything back, you can't put a dent in it, it doesn't take your shape, it doesn't have any smell, all it does is make your nose cold. I like the shine of it—it does look smart, intelligent in an icy way. But I resent its cold impenetrability, its supercilious glare.

It is not surprising that persons holding such a perception of mathematics—as a rigid and impersonal discipline—will avoid engaging in mathematical activities, regardless of their ability! Students' conceptions and expectations, however, can influence their everyday approach to mathematics learning in less obvious yet powerful ways, as the following episodes illustrate.

One of my greatest surprises as a teacher and a researcher occurred at the beginning of an experimental eleventh-grade mathematics course, which I designed and taught in a local alternative high school. I was proud of the plan for my first class, which was to be devoted to the discussion of *why* the outrageous simplification $\frac{1\not{6}}{\not{6}4} = \frac{1}{4}$ gives a correct result—an intriguing problem that requires genuine problem solving and invites reflection about the nature of equations and methods for their solution. I also felt that the class had gone reasonably well, since the students had been attentive, and a few of them had shown considerable creativity in their approach to the problem. Little did I know what lay in store for me! I learned the next day that many of my students, far from bring intrigued by my approach, were actually planning to drop the course! The best I could do (once a first reaction of panic was subdued) was to have a frank talk with the students in our next class. There their reaction became suddenly clear as they expressed their deep concerns through a deceptively simple question: "How can we ever cover the course material if we spend a *whole* class on just *one* problem?" This question revealed the asymmetry in expectations that I as a teacher and they as a class shared about what makes a mathematical activity valuable. Such a discrepancy certainly went a long way in aborting the success of my lesson, regardless of the pedagogical merit of its design.

The experience described above is in no way unique to the introduction of experimental teaching techniques, as shown by the following episode, which occurred in a regular class and dealt with a more *traditional* curriculum content:

> Our class had been discussing the coordinate system, three points along with their coordinates were drawn on the board, and the students were asked to determine if these points were collinear. After the definition of collinearity had been recalled and discussed, there was a long silence broken finally by a student who said that he did not know the formula for collinear. When the explanation was made that no such formula existed, he replied, almost in anger, "How do you expect us to do a problem if we've never been shown how?" The instructor replied, "Think about it." All gave up except one student. He responded (a bit hesitantly) that he did not know how to do the problem *mathematically* but he did think he could do it his own way, and proceeded to reason that if A, B, C were three points (in that order) on a straight line, then the distance from A to B added to the distance from B to C must be the same as the distance from A to C; otherwise they were not collinear points and formed a triangle. Of course, the teacher was delighted with his answer and asked why he did not think it was mathematical. His reply was: "Because I thought of it myself." (Oaks 1987, pp. 3–4)

In this episode, the students' implicit assumption that mathematics consists of a predetermined set of rules and procedures "passed on" by teachers to the next generation did not allow them to consider *thinking on their own* as an appropriate strategy to approach mathematical problems.

As illustrated by these examples, students' conceptions of the nature of mathematics and their expectations with respect to school mathematics can constitute a powerful force operating "behind the scenes" in any mathematics class. Since these beliefs are usually deep-seated and unconscious, it is unfortunately difficult for teachers to access them, and even more so to attempt to modify those conceptions that may seem counterproductive, as revealed by several research studies on people's belief systems (see, for example, Perry 1970; Cooney 1985; Brown and Cooney 1987). Just "telling" the students what mathematics really is and what is expected from them is not likely to do much to resolve the problem.

With the goal of helping teachers deal constructively with their students' views of school mathematics, I have attempted in this article to identify some beliefs that could prove dysfunctional to students' learning of mathematics and to discuss ways to help students gain a better appreciation of the nature of mathematics and consequently develop expectations and behaviors that are more conducive to success.

SOME "DYSFUNCTIONAL" MATHEMATICAL BELIEFS

The results of several recent research studies have shed light on students' views about school mathematics (see, for example, Buerk 1981, 1985; Oaks 1987; Schoenfeld 1985). To "capture" even unconsciously held mathematical beliefs, a variety of methodologies were employed, including a combination of open-ended questionnaires, in-depth interviews, student journals, videotapes of problem-solving sessions, and even the analysis of metaphors of mathematics created by students. These studies have revealed the *existence* of a set of beliefs that could negatively affect mathematics instruction. These beliefs are briefly synthesized below—with some inevitable oversimplifications—with respect to four key categories:

- *The scope of mathematical activity:* Providing the correct answer to given problems, which are always well defined and have exact and predetermined solutions. This applies to the activity of both mathematicians and mathematics students, though the complexity of the problems approached would obviously differ.
- *The nature of mathematical activity:* Appropriately recalling and applying learned procedures to solve given problems.
- *The nature of mathematical knowledge:* In mathematics, everything is either right or wrong; there are no gray areas where personal judgment, taste, or values can play a role. This applies both to the facts and procedures that constitute the body of mathematics and to the results of each individual's mathematical activity.
- *The origin of mathematical knowledge:* Mathematics always existed as a finished product; at best, mathematicians at times discover and reveal some new parts of it, while each generation of students "absorb" the finished products as they are transmitted to them.

In what follows, for brevity, let us refer to the view of mathematics characterized by this set of beliefs as "dualistic." I would like now to support my original claim that the mathematical beliefs identified above may prove dysfunctional for their holders' learning of mathematics.

First of all, these beliefs reflect a limited appreciation of the nature of mathematics. They are oblivious of the struggle and creativity that was required to achieve even what may now be considered the most basic mathematical results. The presence of controversy on some mathematical issues, and in particular on the foundations of the discipline, is not even conceived as possible. Ignored is the role played by value judgments in the creation of new definitions, the choice of an axiomatic system, or the evaluation of alternative proofs for the same theorem. These beliefs also fail to capture the complexity and nonlinearity of mathematical applications.

Even accepting that a dualistic view of mathematics does not reflect the nature of the discipline, one might still wonder whether it really represents a problem for mathematics students, especially at an early stage of schooling. After all, a sense of security may even be derived from perceiving a discipline as perfectly organized and void of ambiguity. And there could be the hope that a student's limited conception of mathematics will modify and expand naturally as more advanced mathematical topics are encountered.

Unfortunately, this may be true at best only for some of the most able students. For the majority of the students, a dualistic view of mathematics is more likely to cause expectations and behaviors leading to anxiety and academic failure. For example, the students' reactions in the episodes reported earlier can now be easily explained, and could even have been predicted, as a direct consequence of these mathematical beliefs. And so are other ineffective student behaviors that usually puzzle mathematics teachers, such as focusing on memorization rather than on conceptual understanding or lacking constructive strategies to cope with learning difficulties. Additional examples of counterproductive expectations and behaviors that could be derived as "corollaries" from a dualistic view of mathematics can be found in figure 21.1.

What can have caused students to develop such mathematical beliefs, since these beliefs do not truly reflect the nature of mathematics and certainly cannot be justified on the basis of facilitating the learning of mathematics? Social stereotypes and stages of intellectual development may certainly play a role in shaping students' conceptions, as suggested by research on world views such as those of Perry (1970) and Belenky et al. (1986). Yet another cause could be the way mathematics is presented in school. We should not be surprised at mathematics students' overwhelming concern with product and answers when the most important measure of academic success is given by the score received on standardized multiple-choice tests taken under considerable time pressures. Nor can we ignore the fact that a

Some Common Misconceptions and Their Explanation in Terms of a Dualistic View of Mathematics

Learning mathematics is a straightforward matter and practice *alone* should "make perfect."

If mathematical activity is equated with applying the appropriate algorithms to given problems, then learning mathematics should only involve taking down notes of the procedures that the teacher gives out, memorizing all the steps in their correct sequence, and practicing on a sufficient number of exercises so as to become able to perform the procedures quickly and without mistakes; and if the results are not satisfactory, all you can do is practice more.

It is no good trying to reason things out on your own.

If you learned successfully, you should be able to do the problem quickly; if instead you did not pay sufficient attention to the teacher in class or did not practice enough, "thinking" alone cannot help remedy these deficiencies; furthermore, since for any problem there is a correct procedure to be applied and you do not know it, you cannot hope to come up with another one on your own.

Staying too much on a problem is a waste of time.

It is difficult to appreciate that by looking closely at *one* problem, you can learn something more general and transferable, when "reasoning it out" is not conceived as appropriate and furthermore there is the perception that mathematical results are disjoint.

You cannot learn from your mistakes.

If there is no connection between a right and a wrong way of doing mathematics, trying to analyze and understand your mistakes is just a waste of time; therefore, if you do something wrong, you should forget about it and start back from scratch to do it right.

Formal mathematics is just a frill.

Proofs, deductions, formal definitions, are not really helpful when it comes to finding and correctly applying the appropriate algorithm to solve a problem; therefore, formal mathematics becomes a ritual that can be performed on the teacher's request but that can be ignored in the context of solving problems.

History and philosophy of mathematics are irrelevant to learning mathematics.

A reasonable conclusion, if mathematics learning is defined in terms of mastering procedures to solve specific problems and such procedures are believed to exist independently of the way they were discovered; consequently, readings and teachers' lectures in these areas are likely to be perceived as a digression and a "luxury," which a student struggling with mathematics cannot afford.

A good teacher should never confuse you.

From the assumption that ambiguity does not exist in mathematics, it follows that the origin of confusion must be a poor presentation; this expectation can easily justify "blaming the teacher" for the learning difficulties experienced and also make students resistant to innovative teaching approaches based on discovery, problem solving, and explorations.

Fig. 21.1

simplistic view of mathematics learning could simply reflect what goes on, almost every day, in most mathematics classes: the teacher introduces a new concept or rule by lecture, applies it in a few simplistic examples, assigns similar exercises for practice to the students for homework, later on goes over these homework results, and finally verifies the students' ability to perform the same task in a test situation.

STRATEGIES TO HELP STUDENTS RECONCEIVE THEIR VIEWS OF MATHEMATICS

The previous analysis of some dysfunctional conceptions of mathematics has made clear that helping students reconceive their views of the discipline should emerge as a major concern for mathematics teaching. At the same time, the challenge of such a task has also been revealed. Because of the very nature of a dualistic and product-focused view of mathematics, we can in fact expect that a direct approach through lectures and readings *about* mathematics will do little good, since students are likely to perceive such initiatives as merely "icing on the cake" and consequently pay little attention to them. Nevertheless, the task is not impossible, and valuable insights and strategies to approach it have been suggested both by research on how conceptions develop (such as Perry 1970; Oaks 1987; Brown and Cooney 1987) and by interventions that actively attempted to affect people's conceptions (such as Buerk 1981; Borasi 1986a, 1988).

All these studies have pointed out the key importance for students to *become aware of, and reflect on, their beliefs, as well as possible alternatives,* since beliefs are more powerful the more they are held unconscious and unquestioned. They have also made us aware of the danger of rhetoric: in abstract discussions on the nature of mathematics and its learning there is too often the risk of cross-talk and misunderstanding because students and teachers may not attribute the same meaning even to key terms such as *understanding* or *mathematical problem.* It is important, therefore, to offer students opportunities to *engage in mathematical activities that can generate doubt* in assumptions taken for granted up to that point and to *personally experience more humanistic aspects of mathematics* of which they may have been unaware.

For example, some cognitive conflict could be stimulated by presenting students with unsolvable mathematical paradoxes and contradictions, for which even "Authority" does not know the answer; inherent limitations, as in attempting to define 0^0; statements that may be right or wrong depending on the context, such as "multiplication is repeated addition." All these experiences, in fact, are in conflict with the dualistic expectation that mathematics is a perfect domain, where everything is either right or wrong, and may consequently lead students to reconsider the validity of such beliefs. In turn, encouraging students to share their procedures to perform a specific mathematical task could shake the belief that the correct result in mathematics can be reached in only one way and pave the way for the acceptance of alternatives in other aspects of mathematics as well. A better appreciation for the role of personal judgment and values in mathematical activity could instead be fostered by engaging students in the study of problem situations (National Council of Teachers of Mathematics 1989), where the solver is required to define the problem more precisely, select relevant information, and evaluate the acceptability of the solution(s) reached. Problem-posing activities, where students are encouraged to generate on their own mathematical questions worthy of study, could also greatly contribute to this goal (Brown and Walter 1983).

An awareness of critical events in the history of mathematics, and of the troubled genesis of some specific topics and results, could well complement the previous experiences by showing how even great mathematicians had to struggle and use considerable ingenuity to produce what is now accepted within the body of mathematical knowledge.

Although the strategies described above can help teachers design activities that may generate healthy doubt in their students' minds, it is also important that the time and opportunity to *reflect on these experiences* be provided. Class discussions are certainly a good means to do so, but there are considerable advantages to complementing them with expressive writing activities. Writing assignments requiring students to report and reflect on their mathematical experiences, perhaps with the help of a few thought-provoking questions by the teacher, will in fact force each individual student to take a stance, can provide more time and leisure to identify, work out, and express satisfactorily one's ideas, and finally will produce a written product that could be exchanged with others and provide a record of development over time (see, for example, Borasi and Rose [in press] and the article by Azzolino in this volume).

It is also important, however, that the activities thus described do not remain isolated episodes within an essentially traditional mathematics curriculum. Rather, throughout the curriculum the students should experience aspects of mathematics and mathematical activity that are consistent with a nondualistic view of the discipline. This requires that potential problems, controversial points, and the possibility of alternative interpretations be continuously highlighted and discussed rather than hidden from students' consideration, even at the risk of occasionally losing some clarity. Students should also be put in a position in which they can engage in the creation of original answers, the generation of new questions, and the evaluation of their own mathematical activity instead of always being immediately furnished with the final and "correct" product, even if this process will require more time and consequently the quantity of mathematical results covered in the curriculum may need to be reduced.

Obviously, the strategies described in this section cannot be seen as simple additions to the existing curriculum and traditional teaching approaches; rather, they require that we radically reconsider the goals, content, and means of mathematics instruction for the 1990s.

CONCLUSIONS

In this article I have examined the nature and consequences of students' holding a dualistic view of mathematics—that is, the belief that mathematics is a collection of disjoint, predetermined, and absolutely correct facts and procedures that are used to solve specific problems and that teachers are supposed to pass on such facts and procedures to students, who in turn will memorize them for later recall and application for the solution of given problems. It has been argued that students holding such a view are likely to be passive learners, focusing on memorization rather than conceptual understanding, ignorant of the need of making personal meaning of the material presented in class, lacking constructive strategies to deal with learning difficulties, and therefore liable to experience limited success in school mathematics.

If this is true in the context of the current curriculum, which still mostly stresses the acquisition of technical facts and skills, it will be true even more so in the future. The new goals for school mathematics in the 1990s, put forth by the National Council of Teachers of Mathematics to better respond to the needs of our ever-changing and increasingly technological world, are in fact even more at odds with a dualistic view of mathematics, since they stress the importance for students of "(1) becoming a mathematical problem solver; (2) learning to communicate mathematically; (3) learning to reason mathematically; (4) valuing mathematics; (5) becoming confident in one's ability to do mathematics" (National Council of Teachers of Mathematics 1989, p. 12).

Overcoming the problem of students' inappropriate conceptions of mathematics should thus become a priority for mathematics instruction in the 1990s. Yet, this article has shown that this is no easy task. We cannot hope to resolve the problem by adding an isolated unit on "The Nature of Mathematics." Rather, teachers will need to create a variety of learning situations throughout the curriculum, which will lead students to become aware of, and question, their perceptions of mathematics as they experience and reflect on humanistic aspects of mathematics and its learning. It is hoped that this article has provided the stimulus and encouragement, as well as some concrete suggestions, to engage in such a challenging and worthwhile enterprise.

REFERENCES

Belenky, Mary Field, Blythe McVicker Clinchy, Nancy Rule Goldberger, and Jill Mattuck Tarule. *Women's Ways of Knowing.* New York: Basic Books, 1986.

Borasi, Raffaella. "On the Educational Uses of Mathematical Errors." Ph.D. diss., State University of New York at Buffalo, 1986a.

———. "Behind the Scenes." *Mathematics Teaching* 117 (December 1986b): 38–39.

———. "Using Errors as Springboards to Inquire into the Nature of Mathematical Definitions: A Teaching Experiment." Preliminary report on project no. MDR8651528 sent to the National Science Foundation, 1988.

Borasi, Raffaella, and Barbara Rose. "Journal Writing and Mathematics Instruction." *Educational Studies in Mathematics,* in press.

Brown, Stephen I., and Thomas Cooney. "Stalking the Dualism between Theory and Practice." In *Second Conference on Systematic Cooperation between Theory and Practice in Mathematics Education, Part 1: Report,* edited by P. F. L. Verstappen, pp. 21–40. Lochem, The Netherlands: National Institute for Curriculum Development, 1988.

Brown, Stephen I., and Marion I. Walter. *The Art of Problem Posing.* Philadelphia: Franklin Institute Press, 1983.

Buerk, Dorothy. "Changing the Conception of Mathematical Knowledge in Intellectually Able, Math Avoidant Women." Ph.D. diss., State University of New York at Buffalo, 1981.

———. "The Voices of Women Making Meaning in Mathematics." *Journal of Education* 167 (3)(1985): 59–70.

Cooney, Thomas J. "A Beginning Teacher's View of Problem Solving." *Journal for Research in Mathematics Education* 16 (November 1985): 324–36.

National Council of Teachers of Mathematics. *Curriculum and Evaluation Standards for School Mathematics.* Reston, Va.: The Council, 1989.

Oaks, Ann. "The Effect of the Interaction of Mathematics and Affective Constructs on College Students in Remedial Mathematics," Ph.D. diss., University of Rochester, 1987.

Perry, William G., Jr. *Intellectual and Ethical Development in College Years: A Scheme.* New York: Holt, Rinehart & Winston, 1970.

Schoenfeld, Alan. *Mathematical Problem Solving.* Orlando, Fla.: Academic Press, 1985.

22

Contextualization and Mathematics for All

Claude Janvier

CALLS for reform in school mathematics (Cockcroft 1982; National Research Council 1989; National Council of Teachers of Mathematics 1989) have focused on "mathematics as sense making" and on the importance for all students in grades K–12 to study a common core of broadly useful mathematics. An examination of particular ways in which mathematics is used suggests that changes in both curriculum design and teaching are needed.

Recent studies such as Carraher, Carraher, and Schliemann (1985, 1986) and Lave, Murtaugh, and de la Rocha (1984) have investigated the way arithmetic is used by such persons as supermarket shoppers, young market merchants in developing countries, illiterate carpenters, bookkeepers, and lottery ticket sellers. Such an investigation is particularly difficult, as suggested in the following excerpt from the Cockcroft report (1982, p. 18):

> Many jobs require the employee to make explicit use of mathematics—for instance, to measure, to calculate dimensions from a drawing, to work out costs and discounts. . . . However, even when mathematics is being used, frequent repetition and increasing familiarity with a task may mean that it may cease to be thought of as mathematics and become an almost automatic part of the job. A remark which was overheard (in surveys)—"that's not mathematics, it's common sense"—is an illustration of this.

Taking into account this basic difficulty, these investigators have consequently adopted a combination of observation techniques and interview procedures that do not simply rely on people's impressions or beliefs.

Their conclusions converge. They all show that arithmetic is effectively used and, moreover, used very efficiently. They even at times make calculations that are more complex than the ones they learned at school. These calculations are usually error-free. For example, for the supermarket shoppers, only 2 percent of their computations were inaccurate. For the young market merchants, the success rate was even better. But as far as school arithmetic is concerned, results obtained by those people on subsequent tests given by the investigators and at the same level of difficulty were appallingly low. The supermarket consumers scored only 59 percent on a standard test of basic arithmetic operations and 57 percent on a test on ratio comparisons designed as typical of supermarket computations. As for the young merchants, a week later they took a test that included the items that they had successfully completed during their transactions in the market place, but their results were statistically lower.

Making a living out of those "computations" imposes, so to speak, techniques that will ensure success and with which one will feel confident. The major discrepancy observed can be mainly attributed to the fact that the computation methods used in real-life situations are different from the ones typically introduced in school. Too, the symbolic modes in which the calculations are expressed are different—for instance, mental calculations in the market place and written tests for the school arithmetic test.

The crucial point is that the computation methods people use are determined by the situation or the context in which they are performed. For instance, in the problem *Three coconuts at 80¢ each will cost* (as the child says) *80, 160, 240: $2.40,* we note that the child performs a repeated addition articulating or punctuating the counting process on the coconuts. For calculating *How much are 12 melons at 50 cruzeiros,* a youngster more or less mentally takes them two by two and counts as follows: "100, 200, 300, 400, 500, 600." Sometimes the procedure can become more tricky: 10 times 35 can sometimes be carried out in the head as 105 (3 times 35) + 105 + 105 (which makes) . . . 315 + 35 (since there is one missing) . . . 350. Note that this is equivalent to (3 + 3 + 3) times 35 plus 1 times 35: a particular use of the distributive law of multiplication on addition. The well-known school technique of adding a 0 seems to be suspicious.

In ratio and proportion problems, going back to the value of the unit is not systematically used. Note that the value of unit is not always given since in the market place, two, three, or four items may be offered for an easy multiple (e.g., 10, 50, or 100) of the money unit. Let us give an example: If 2 melons cost $5.00, how much do 6 melons cost? In this example, the value of 1 melon ($2.50) will probably not be evaluated. Buyers will consider that they want three times as many melons and that consequently they will pay three times as much: $15.00. Some combinations of numbers will make such a procedure often difficult, but it is by far preferred to the other one that consists in finding the rate, namely, the price of a unit or of the unity. This has already been pointed out by Freudenthal (1983), Vergnaud (1983), and others.

THE NOTION OF CONTEXTUALIZATION

Those unexpected observations deserve explanation. Apparently, the tests done with pencil and paper automatically bring about a more or less memorized algorithm, whereas the "in the action" problem is tackled more freely. Taking the context into consideration implies that numbers are processed in the operations without losing their situational connotations. In other words, computations are not made with abstract written numbers but rather on quantities, magnitudes, or measures. For instance, money problems will give rise to operations that can be based on the relations existing between the various coins and bank notes. This is one of the ways through which the context plays an active role in supporting the reasoning toward the solution.

This is even true of the arithmetic operations that would convey meanings that are inspired from the elements of the context. Multiplication problems may sometimes be treated as addition problems in the way the melons (either actual or as mental images) are used to support the reasoning. Division problems are "contextually adjusted." For example, the question How much for one? is often avoided, since combining the units (to obtain a rate) cannot be supported easily by a contextual entity derived from the combination of observed numerical entities.

In fact, since each context involves particular features, it is normal to expect that certain aspects of the counting procedures or of the reasoning processes will be emphasized in view of the actual features that are being stressed. So far, the argument has been mainly based on a fair amount of research results and has concerned only arithmetic. However, my personal experience and research have provided similar evidence for more advanced mathematics.

MATHEMATICS AND CONTEXTUALIZATION

Indeed, if ordinary citizens use arithmetic their own way (as shown above), it seems reasonable to assume that scientists, engineers, and technicians will use mathematics similarly. Do they rely on some contextual cues when solving their problems? Can it be likewise when they cope with equations or when they make use of functions, integrals, or derivatives?

Making use of mathematics is generally associated with setting up equations or formulas and solving them. At a more basic level, as we have seen in the examples above, it only implies establishing a simple arithmetic operation or a relation of proportionality. Do engineers or scientists go beyond this level? Past experience with science lecturers at the university level tends to convince me that they do. The next example will illustrate my point.

Solving an Electricity Circuit

The end of most chapters on elementary electricity present problems that ask students to find the resistance of resistor R_2, given, for example, that the current is 3 amps and that the battery voltage is 120 volts (fig. 22.1). The internal resistance of the battery is negligible.

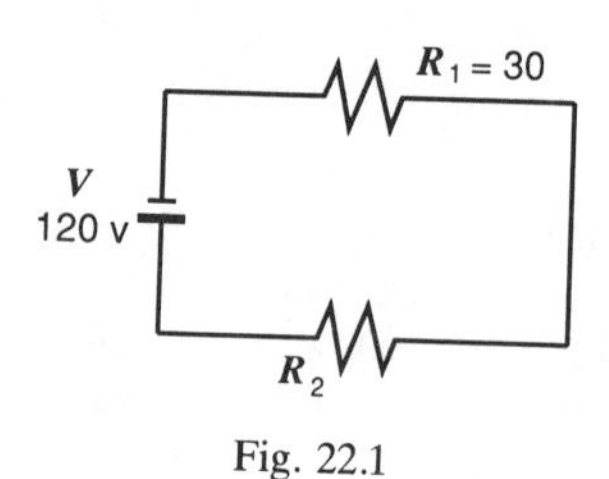

Fig. 22.1

In solving the problem, students are expected to set up equations. To start with, it needs little inspiration to write down $V = V_1 + V_2$. Subsequently, they are expected to relate the voltage with the resistance and the current and write $V = R \cdot i$ and $V = R_1 \cdot i + R_2 \cdot i$. The last equation can be derived in many ways. The main goal is the substitution of values, as indicated below:

$$120 = 3 \cdot 10 + 3R_2$$
$$120 - 30 = 3R_2$$
$$R_2 = 90/3 = 30$$

The point to emphasize here is that a lot of manipulations are performed without resorting to the circuit diagram. It would be interesting to know more precisely what the real contribution of this diagram is to solving the problem. But in any event, the vast majority of physics teachers would not write down equations when it comes to solving such a simple electricity exercise. The equations would be replaced by the diagram itself, which would be used to combine the quantitative relations between the variables involved. The context here can be considered as a mixture of the well-known basic laws of current and the diagram, which guide and support the reasoning. The equations implicitly used by teachers (and having a strong schematic content) will differ from those of the students, which are mostly "algebraically inspired" in that they are totally dependent on the allowed transformations, for example, putting the unknown on one side. In contrast, when our laboratory technician inspected the circuit, a total resistance came to his mind going with a voltage drop in two steps. Indeed, 3 amps for a voltage drop of 120 volts "requires" 40 Ω. The split 30 Ω and 10 Ω (which gives 40 Ω) appears quite clearly on the diagram. In short, the fundamental relations that are conveyed by the equation were immediately visible on the diagram. In this sense, the diagram replaced the equation and was most likely used as the mental images or concrete objects that help the young market merchants.

Exponential Growth

The following riddle is well known for having fooled many people:

A 256 m^2 pond contains water lilies. As they grow, water lilies double in area each day. If the pond is fully covered after 7 days, when was it half covered?

If this problem is assigned during lessons on exponential functions, it will give rise to a lot of juggling around on equations with the aim of obtaining a formula of the type $p(x) = k2^x$. Finally, the students who get to $2^x = 64$ will most probably use logarithms or their calculator to get an answer!

We note that as soon as the problem is identified with a mathematics exercise, a search for the equation and its solution is triggered. At first, the parameters of the equations that are linked to the situation are called into play in the process of solving, but the mathematical rules for dealing with the equations quickly lure the students away from the context, which is finally only (and slightly) referred to when data are needed. However, if one considers the expansion of a large spot that progressively covers a closed curve, it is likely (but not necessarily easy) that this mental image helps to associate doubling with halving and being half covered. Note that the solution can be arrived at through an "undoing" or a playing back of the water lilies' growth. The context must be tapped at the visual level, although one can imagine some verbal patterns being resourcefully exploited as well.

These two examples illustrate the effects of teaching practices that focus the solution process on setting up equations and transforming them. They demonstrate how efficient problem solvers rely on the context of the problem and the fundamental relationships implicit in the problem to drive their reasoning rather than on transforming the relationships to equations and then manipulating the equations. In fact, it is their flexible mastery of basic principles that enables them to work with several symbolic modes. It is this context that is so often neglected or diminished in solving real problems in school mathematics.

CONTEXT AND SCHOOL MATHEMATICS

The previous examples highlight various beliefs about doing mathematics and using mathematics. Actually, doing mathematics is too often identified with searching for the unique, most powerful, abstract, and general solution. Students "live" with the impression that there is a canonical way to perform an operation and state a problem—methods that have the property of being adaptable to all particular cases. This is why equations are often preferred to other modes of solving problems, such as using diagrams, tables, or numerical relations. Certain representations are less "acceptable" than others.

Moreover, mathematics also involves performing operations with a certain degree of automaticity that is void of contextual details. Using mathematics is clearly associated at some point with forgetting about the meaning and basic quantitative relations concerned. In fact, using mathematics is synonymous with applying mathematics, an expression that I will now try to characterize.

Most schools are run according to an implicit agenda in which mathematics is first learned in the mathematics lessons and then *applied* in the science lessons and everywhere else. This epistemological tradition tacitly shared by the scientific community (including mathematicians and scientists) confers on mathematics a preponderant role with respect to science in general. Once a problem is mathematically formulated, the answer can be obtained through automatic processing that is devoid of context.

There is consequently no surprise to the fact that applying mathematics in science is identified with writing down equations for the end-of-chapter problems. The consequences are that the chapter equations are learned by heart and recalled more or less on the basis of the factors being present in the text's exercise statements. Most of the time, the modelization process through which the phenomena are reduced into equations appears only in the teacher's exposition. Consequently, the underlying analysis through which contextual elements are associated to mathematical relations is rarely achieved or realized by the students. It is quickly understood by the students that it is preferable to forget about the situational meaning of the variables, operations, or rules. In fact, for most cases the mathematical tricks should do the job.

In the actual and implicit relations between mathematics and science, the contextual content of the mathematics that is the quantitative aspect of science is at first absent and is expected to arise through applications. In other words, the contextual richness or depth is expected to be added to the mathematical concepts within the application exercises. In fact, the science teachers as the "appliers" make use of notions that are, so to speak, bestowed by the mathematics teachers. But even though this epistemological perspective is never challenged, the day-to-day "life" in school is not so simple. It frequently happens, for instance, that science teachers express concerns about the fact that the mathematics teachers have not quite "prepared" the "right" object. It is well known that students have to deal with mathematical ideas presented by their mathematics teachers and their science teachers—for example, vectors and logarithmic functions. It seems that the mathematical (axiomatic or otherwise) approach does not seem to fit with the necessity to establish quantitative relations between science concepts. Thus, the notion of applying mathematics has to be reexamined.

REPRESENTATIONS AND CONTEXTUALIZATION

The examples presented so far stress the fact that the support of the context makes the equations unnecessary for solving the problems. The solution methods can be inspired from the context and represented through several representations. Indeed, the basic connections can be singled out through the use of alternative symbolizations: diagrams, verbal descriptions, or mental images, for example. These representations allow the essential features to be singled out and are used as a guide toward the solution. The rationale is that a certain proximity with the situation and with the way the variables are articulated contributes to the solution process. The notion of contextualization has been introduced to convey the idea that the context is kept in play. Mathematical ideas that are processed have kept some concrete connotations, or in other words, they are not entirely stripped of their situational content.

Consequences for Teaching

The Cockcroft report agrees with most of what has been presented here, even though the cases are less documented. There exists a contextual mathematics that is not as such developed at school but that is used in the workplace. I contend that it is not only developed in the working place but that it is built on some basis that is developed at school. The problem seems to be that according to criticisms, this basis is too often inappropriate. As the Cockcroft report (1982, p. 24) states:

> It is important that the mathematical *foundation* which has been provided in the classroom should be such as to enable *competence* in *particular applications* to *develop* within a reasonably short time once the necessary employment situation is encountered. [italics added]

I have deliberately emphasized the words that can now be differently and more deeply appreciated. Indeed, the aim of enabling competence in particular applications to develop must be reassessed. In fact, comments made so far in this article have shown that using mathematics is more akin to contextualizing mathematics than to applying it. In other words, the actual practice of many "users" is not simply a particular case of a general method learned at school. Consequently, the whole notion of mathematical foundation has to be revised.

It has been strongly argued that foundation cannot be equated with some general mathematical knowledge that can be applied. It is not so much a matter of examining the content or the thinking process as the way the foundation is developed. In fact, schools should stress diverse contextual explorations of the foundations. There is a need to introduce contexts in the classrooms.

Growth and Contextualization

Almost any year, students can spend many hours on the notion of growth. This general idea involves the children themselves, pet animals, life in general, the city and country populations, and cost of living. Remember that as far as possible the objective is to make sure that the representations are such that the context is kept active in the reasoning process (as a meaningful guide). There are numerous contexts in which the notion of growth fits naturally: changing furniture, buying clothes, defining maturity, the need for a new school, or real estate development. According to the school level, several questions could be tackled using various representations: tables, growth curves, verbal description, numerical rates, and equations. It is understood that teachers must be very open-minded to accept youngsters' proposals and be aware of the multiplicity of suggestions that can arise. There is no need to insist that a lot can be done with graphs and tables of growth that implicitly contain the notions of rate and slope. For instance, when the graphs are plotted with color pens on different transparencies, comparisons by superimposition allow implicit uses of rates.

The notion of rate of change is central, and it must be kept contextually significant. It means that at the secondary level, equations and formulas must come at the right moment, with the contextual meaning of each parameter being made properly explicit. After the additive growths come the multiplicative ones and a wide variety that can be graphically illustrated.

A lot can be achieved with exponential (multiplicative) growth by using the constant function on an ordinary calculator. The successive values of the geometric sequence $k \cdot a^n$, $n \in N$, can be obtained quickly. For instance, in a problem on compound interest, if k is the initial amount and $a = 1 + i$ (i being the interest rate), any change of 1 in the n's represents a change of 1 basic time period, and pressing the "=" button gives the amount for the following year.

We noted earlier that the "diagram" context for the electric circuit was absent as well as the image of the water lilies growing exponentially. In the present example, the many explorations organized with calculators are aimed at inserting the notions of growth into the context of lending or borrowing money without trying to be exhaustive.

Contextual Growth and Equations

Some problems of exponential growth (or decay) do not involve an intrinsic rate (of growth) with a basic time period over which the rate applies. This is the case of the discharge of a condenser or an atomic disintegration (decay) of a chemical element. The rates of change define differential equations. The variations are described in the solutions with the help of an exponential function whose unique base is $e = 2.7182. . . .$ But mathematical reasons are not always very contextual, as the following excerpt of *The Limits to Growth* by Meadows et al. (1972, p. 30) explains:

> It is useful to think of exponential growth in terms of doubling time, or the time it takes a growing quantity to double in size. . . . A sum of money left in a bank at 7% will double in 10 years.

Multiplicative growth processes can always be represented by an equation of the form $y = k \cdot 2^{x/t_0}$ provided t_0 is naturally adjusted. Actually, the adjustment is hardly necessary, since t_0 is precisely the doubling time, or the half-life, for decay. This means that for the 7 percent growth considered above, we have the identity: $1.07^x = 2^{x/10}$ (doubling every 10 years as it was stated, $t_0 = 10$). In fact, it is intriguing that scientists and mathematicians keep using the base e for exponentiation and keep asking for the doubling time or the half-life period when a proper equation would include this crucial factor as a parameter, namely, the equation of the form $y = k \cdot 2^{x/t_0}$.

The Pedagogical Framework

The example of growth has shown how representational means keep mathematics close to the context and to the reasoning used to solve the problem. Progressive familiarity with the context should then be the goal. Both mathematics and the context analysis should move slowly. In fact, rushing through the context to get to the more powerful and faster techniques only short-circuits one's familiarization with the basic context-mathematics relations. In other words, familiarity with the process of easy modelization can be derived only on the basis of a lengthy involvement. The teacher must also allow a certain flexibility in the tools used by the students. The question is not to apply something general but to explore and become acquainted with the basic relations as they are determined by the various contexts.

The nature of the skills or abilities that have to be developed (process more than content) leads directly to an integration of the scientific disciplines through problem solving. It means, in fact, that a real problem-solving approach should be adopted in which the students are confronted with challenges concerning themselves. Of course, topics will be discussed that do not necessarily belong to the school curriculum, and science questions will be treated as well. In fact, students in the classroom should be introduced to the mental processes of combining mathematics and context.

Realizing these interdisciplinary conditions is much easier for primary level teachers than for teachers at the secondary level, where the complications of curriculum, time schedule, and tradition exist. As an intermediate step toward integration, one can think of better correlations between teachers. Under this hypothesis, more concern for contexts should be found in mathematics lessons, and a different treatment of mathematics is expected in science classes. As far as the attitudes of teachers are concerned, science teachers should show more independence vis-à-vis the "content" of the mathematics teachers, and more flexibility should be expected from the mathematics teachers.

To respect this plan, lessons must be looked at differently. They are not predetermined. Since contexts appear as open issues, the path taken by the children can only be slightly guided. Some degree of freedom is bound to exist. Teachers need to be supported in view of this modified role. The textbook consequently becomes simply a guide to help organize this difficult form of teaching. It must be done on the basis of challenges that concern the students directly, since meaning should be considered as a personal construction that has to be socially adjusted. The work in the classroom must be organized accordingly. The usual magistral expositions should be enriched with techniques such as group works, debates, children's presentations, and school surveys. A laboratory should be shared by mathematics teachers in which contextual measuring, estimating, and prediction is made possible.

Familiarization is not basically achieved through a discovery learning approach. A balance should be reached between what is proposed by a group of children and then put forward to the class and the suggestions coming from the teachers (often inspired from what was done the previous years).

Where Is (or What Is) Mathematics, Then?

This analysis suggests that mathematics should be considered a necessary systematization of those contextualized notions and processes that are developed. Systematization could or should come once in a while during the year as a need to converge on a common agreement. The whole body of contextualized mathematics would become the context from which the construction or the organization of mathematics would take place. And this would be the same thing for science. The systematization would be derived from a familiarity based on primitive but efficient representations. In fact, there would exist in the long run a form of fuzzy discipline.

All this is mere utopia unless school rules are changed. A new balance in the timetable and a truly original way to deal with classroom and program organization has to be invented. It should allow at the same time the exploration of contexts and some form of convergence toward structure. The process is not continuous in the sense that a critical mass of teachers within a school is necessary to counteract the inertia inherent in most schools.

CONCLUSION

Results of research suggest that the notion of "applying mathematics" has to be reexamined. Most people have no notion of abstract mathematics from which particular applications are derived. The concepts used in problem-solving situations seem to keep some connotations related to the contexts. Introducing the context in the classroom is the major consequence for teaching mathematics. To accomplish this both the content and the means by which it is taught need to be revised. An interdisciplinary approach based on real problem solving is consistent with the need to develop in the students the skills related to using "mathematics in context."

Teachers must base their actions on an adequate knowledge of the reasoning processes used in mathematics, on their ability to detect these processes in children, and on their capacity to organize the classroom in such a way as to promote active learning. This suggests that their training must be designed accordingly without losing sight of the fact that a major confrontation has to be expected. Strong beliefs must be dislodged. Greater cooperation should be advocated between science and mathematics teachers.

Finally, the agenda outlined can be realized only if special attention is paid to the role of evaluation (testing) as a determining element in school life. Indeed, since teaching is geared to the examination system, any move toward introducing contexts in the classroom will slowly reinstall a content-oriented curriculum that is easily testable.

REFERENCES

Carraher, David W., Terezinha N. Carraher, and Analucia D. Schliemann. "Having a Feel for Calculations." In *Mathematics for All,* Science and Technology Education, Document Series No. 20, edited by Peter Damerow, Mervyn E. Dunkley, Bienvenido F. Nebres, and Bevan Werry. Paris: Unesco, 1986.

Carraher, Terezinha N., David W. Carraher, and Analucia D. Schliemann. "Mathematics in the Streets and in Schools." *British Journal of Developmental Psychology* 3 (1985): 21–29.

Cockcroft, W. H. *Mathematics Counts.* Report of the Committee of Inquiry into the Teaching of Mathematics in Schools. London: Her Majesty's Stationery Office, 1982.

Freudenthal, Hans. *Didactical Phenomenology of Mathematical Structures.* Dordrecht, Netherlands: D. Reidel Publishing Co., 1983.

Lave, Jean, Michael Murtaugh, and O. de la Rocha. "The Dialectic of Arithmetic in Grocery Shopping." In *Everyday Cognition: Its Development in Social Context,* edited by Barbara Rogoff and Jean Lave, pp. 67-94. Cambridge, Mass.: Harvard University Press, 1984.

Meadows, Donella H., Dennis L. Meadows, Jorgen Randers, and William W. Behrens III. *The Limits to Growth.* London: Earth Island, 1972.

National Council of Teachers of Mathematics. *Curriculum and Evaluation Standards for School Mathematics.* Reston, Va.: The Council, 1989.

National Research Council. *Everybody Counts: A Report to the Nation on the Future of Mathematics Education.* Washington, D.C.: National Academy Press, 1989.

Vergnaud, Gerard. "Multiplicative Structures." In *Acquisition of Mathematics Concepts and Processes,* edited by Richard Lesh and Marsha Landau, pp. 127–74. New York: Academic Press, 1983.

23

Computer-enhanced Algebra: New Roles and Challenges for Teachers and Students

M. Kathleen Heid
Charlene Sheets
Mary Ann Matras

UNLIKE their predecessors of earlier decades, the mathematics students and teachers of the early 1990s will have access to a vast array of powerful computing technology. The new technology brings with it an equally impressive range of challenges for the teachers and learners of mathematics. The first challenge is the creation and adoption of a curriculum that openly builds on available computing power. Remaining challenges deal with the ways in which the roles of teachers and students will change in the implementation of such curricula.

One response to the first challenge has been a computer-based algebra curriculum, *Algebra with Computers* (Fey 1986), that sought—

- to develop, in a computer-rich environment, an effective understanding of algebraic concepts and the ability to solve problems prior to mastery of the conventional collection of technical skills for algebraic manipulation;
- to make the fundamental mathematical concept of "function" a central organizing theme for theory, technique, and problem solving in algebra.

Teachers at two schools, one (with a 70 percent minority population) in the inner suburbs of a large city and the other (with a 1 percent minority population) in the rural environs of a large university, conducted extensive field tests of the *Algebra with Computers* curriculum with ninth-grade algebra classes over a two-year period. In this article we shall share some of the insights about new teacher and student roles, challenges, and responsibilities that we gained by observing these teachers and students (Heid et al. 1988).

A NEW CONCEPTION OF ALGEBRA USING COMPUTERS

The *Algebra with Computers* curriculum differed from the conventional approach to first-year algebra in both content and form. The problem-solving and mathematical-modeling goals of this curriculum distinguished it from the traditional algebra curriculum. The core of the curriculum's content was the exploration of mathematical representations of realistic situations. Students viewed these situations through graphs, tables of values, and symbols and worked back and forth among these representations as they solved problems within a variety of realistic contexts. The curriculum engaged students extensively in cooperative problem solving at the computer. It encouraged students to use the computer as a tool to explore mathematical ideas. It is this "tool-assisted" problem solving that was at the heart of changes in teacher and student roles.

NEW ROLES AND RESPONSIBILITIES FOR TEACHERS

In the implementation of computer-based laboratory explorations, the teacher must become a technical assistant, a collaborator, and a facilitator. In both preparatory and follow-up classroom work, the teacher will need refined skills as a discussion leader and as a catalyst for self-directed student learning. Although one might argue that these roles are hardly new but rather quite desirable in the computer-free classroom, they become not only desirable but necessary in the implementation of computer-based problem-solving curricula.

The Teacher as Technical Assistant

Teachers have long served as "trainers" to students who are learning to execute computational algorithms. Most often, this trainer role takes the form of the teacher demonstrating the targeted skill with the expectation of accurate modeling by the student. When there is time (there always seems to be the need), the "teacher as trainer" becomes the "teacher as technical assistant." As technical assistant, the teacher works individually with the student helping to debug erred procedures. Dependably, the student's written work serves as a data base for clues to the source of the errors.

In computer-assisted problem solving, the teacher's role as technical assistant is expanded and accentuated. No longer is the needed domain of expertise confined to the execution of paper-and-pencil algorithms. The teacher must now provide assistance with the operation of the hardware and with the techniques of using the software as well as with the problem solving itself. As technical assistant in the computer laboratory, the teacher not only must respond to a greater range of calls for help but also must be able to identify the nature of the problems that arise. When, for example, a student enlists the teacher's help because he or she cannot produce a correct graph, the teacher-diagnostician must determine whether the cause is electrical failure, incorrect adjustment of the monitor, incorrect entry of the function rule, incorrect choice of scaling, or any of a variety of other electronic or conceptual problems. It is no longer sufficient for the teacher to have a box of "quick fixes" to dispense when an algorithm goes awry. More so than in the traditional classroom the teacher must become a true problem solver.

The Teacher as Collaborator

A major goal of the school mathematics curriculum is for students to make and test their own conjectures about the relationships between quantities (National Council of Teachers of Mathematics 1989, p. 84). Unfortunately, reasonable conjectures and viable generalizations are far too infrequently generated by students in the traditional classroom. Students seldom feel comfortable enough with the symbols on which they are operating to venture conjectures about relationships among the quantities represented by the symbols. Those who do propose their own generalizations are frequently in error, and the task of evaluating the truth of a student-proposed generalization is often an easy one for the teacher.

In tool-assisted problem solving, students can generate stacks of data that give rise to an array of conjectures, many of which are correct. The teacher is quickly cast in the role of consultant or collaborator, working with students to determine the truth of their conjectures. One example (Sheets and Heid forthcoming) of this collaboration occurred in the first few weeks of the implementation of the computer-based algebra course. Pairs of students were engaging in a computer-based "guess my rule" exploration. Although most of the rules happened to be linear, neither the teacher nor the students were aware of this pattern ahead of time. Besides, students had not yet studied formal algebraic rules and certainly had not discussed the notion that linear functions were of the form $f(x) = ax + b$. Typical computer output for the hidden function rule, OUTPUT = 4 × INPUT − 1, appears below:

INPUT	OUTPUT	CHECK
2	7	
3	11	
4	15	yes
10	23	no

Two students were having unusual success in determining answers, and they shared their method with the teacher. The students had determined that they could find a rule by multiplying the INPUT value by some small positive integer and then adding a positive integer to the resulting product. The teacher verified that this technique worked for the first few examples, then spread word of the technique among the other pairs of computer-lab partners. Later, a pair of students who had adopted the new strategy pointed out to the teacher that it did not work on some of the tables (for example, OUTPUT = INPUT^2 − 3). The teacher quickly adjusted the advice she was offering. Because the tasks have a variety of unpredicted solution paths, teachers and students in the computer laboratories become natural collaborators in the pursuit of those solutions.

The Teacher as Facilitator and Catalyst

A small-group structure emerges naturally in the problem-solving computer laboratory and calls for its own specialized set of teaching skills. With the regular use of computer-lab assignments designed for group problem solving, students must learn to work together effectively and teachers must refine their skills at facilitating the group process. Unlike some small-group working arrangements, student responsibilities at computer stations are ideally split, with one student taking responsibility for keyboarding, the other for checking input, and both sharing responsibility for the overall scheme of the work. Although working styles in many groups will evolve as students gain experience in working with each other, some pairs will have considerable difficulty in negotiating this division of responsibilities, and one of the partners may consistently be dominated by the other. Without interfering with the acceptable and natural evolution in a group's working style, the teacher needs to be facile at assessing the progress of students at computer-lab stations and at encouraging working arrangements that will enhance the process for each pair or group.

In a computer-based problem-solving curriculum, the teacher must serve as facilitator and catalyst not only during small-group computer-laboratory explorations but also during whole-group classroom discussions. A teacher using this type of curriculum frequently begins class by describing a real-life quantitative situation with the goal of inducing students to produce a viable mathematical model of the situation. The realistic nature of the problems naturally leads the discussion in a variety of directions as students suggest and argue the merits of several different mathematical models. Equally wide-ranging are the discussions precipitated by student-generated solution techniques (numerical, graphical, and symbolic). Having fueled the discussion, the teacher must then call on his or her expertise as a facilitator and help students draw the discussion to an appropriate close. The instructional tasks of catalyzing student involvement, facilitating discussion, and gently guiding somewhat unpredictable class discussions to a meaningful close make the computer-enriched problem-solving classroom a more challenging, and rewarding, milieu for teachers.

Responsibilities and Challenges in Evaluating Student Learning

Evaluating student learning in a computer-enriched problem-solving course differs from traditional assessment in major ways. First, as the computer is integrated into the school mathematics program, it becomes more natural to test the wide range of mathematical abilities and dispositions as recommended by the mathematics education leadership (National Council of Teachers of Mathematics 1989). That assessment can take place informally in the regular instructional setting, in a formal testing situation (with computing devices available to test takers), or in one-on-one interviews with samples of students. With the computer available during testing situations, for example, it becomes easier to create questions that do not depend on traditional by-hand skills and test student abilities to identify correct procedures, to analyze problem situations, to use a variety of mathematical representations, and to make and test conjectures. With a greater emphasis on class discussion of problems, teachers have more of an opportunity to assess students' willingness to ask "What if . . .?" or to explain and argue for their viewpoints on mathematical issues. Interviews have long been an excellent tool for probing student understanding, and a computer-enriched problem-solving curriculum suggests a myriad of new ways to deepen and broaden the foci of these interviews.

Second, in a computer-enriched problem-solving environment, student learning is more visible. As students discuss and solve problems at the computer, teachers see more of the problem-solving process and are then confronted in new ways with assessing it.

Third, the shift in goals away from the execution of routine procedures and toward the development of mathematical modeling and problem-solving abilities demands new objectives for evaluation. Instead of being tested on a series of by-hand symbolic-manipulation skills, students in these curricula will be tested on other aspects of the problem-solving process (problem formulation, the selection of strategy, and the interpretation and extension of results). For example, students could be tested on these new skills in the context of the following situation:

> Chuck and Jeanne live in the Snow Belt along Lake Erie and run C&J Snow Removal, a large snow-plowing business. They notice that the number of driveways they can count on plowing each winter month seems to depend on the price they charge for plowing one driveway. One year they decide to analyze the data from a previous year so that they can predict their earnings. Here are the data they want to analyze:

Price charged for each driveway	Number of customers
\$10	6800
\$20	5100
\$30	3400
\$40	1700

To test their ability to formulate problems, students could be asked to generate a rule that best approximates the data (using a curve-fitting program or their knowledge of linear functions, they will find that $n(p) = 8500 - 170p$ works well). They could be asked to comment on other factors that might influence the number of customers and to suggest ways to collect data on the influence of these factors. To test their ability to choose, use, and compare a variety of strategies and representations, students could be asked to select an appropriate representation (numerical, graphical, or symbolic) to answer questions about the situation: "If C&J charge \$27 for each driveway, how many customers can they expect? How many customers will they lose by charging an additional \$2 a driveway?" They could be asked to compare several strategies, citing advantages and disadvantages of each. Finally, to test their ability to extend and interpret computer results, students could be asked to explain what the "170" in the rule $n(p) = 8500 - 170p$ tells about the snow-plowing situation. They could also be asked to graph the revenue curve and to interpret how the shape of the revenue function relates to the snow-plowing business. The computer makes possible and desirable a new array of mathematical modeling questions.

Responsibilities and Challenges in Time Allocation

When computer-based problem solving becomes an integral part of a school mathematics course, issues of time allocation in both the classroom and in the computer laboratory become paramount. Such courses encourage an open-ended exploration of real-world problems, and the length and content of class discussions are often unpredictable. At least two instructional issues arise in the context of this type of classroom discussion. First, teachers find it difficult to know when and how to draw these discussions to a close. Second, when the exact content of classroom discussions is unpredictable, student absenteeism seems to present greater than usual difficulties. Teachers using this computer-based algebra course are gradually refining their skills at conducting classroom discussions, highlighting important points as they arise in the discussion, and closing classes with any major points that the discussion may have missed. They are encouraging students who miss class to conduct minidiscussions on the material with fellow students. Although such interactions will not duplicate the missed class discussions, they are likely to approximate their flavor. Finally, teachers are also beginning to realize that, because the acquisition of concepts and processes in a problem-solving curriculum develops gradually and cumulatively, students who have been absent can still continue to participate meaningfully in subsequent class discussions.

For teachers committed to a computer-enriched problem-solving curriculum, problems with time allocation also affect work in the computer lab. How and when should they signal an end to students who are engaged in computer-based explorations? How should they handle students working at different paces through the lab-exploration material? What should their minimal expectations be for the completion of lab activities? Their usual formulas for introduction/class discussion/guided practice seemed to fall flat when they tried to apply them in the computer lab. If the computer labs are an "add on" to regular classroom instruction, the differing pace and direction of student work in the lab may not substantially affect the general progress of the course. Students who do not finish a particular lab can arrange to complete it at some other convenient time, and teachers can proceed with the course material. If, however, the lab experiences are an integral part of the course development, teachers need to find ways to assure that all students acquire the vital core of shared understanding in an environment that encourages such variation in rate and direction. Unlike assignments in traditional courses, computer-lab explorations cannot be sent home for completion, hence difficulties arise for students who work at a slower pace or who are absent from class.

Teachers working with the *Algebra with Computers* curriculum have begun to formulate answers on how to address these problems. A major feature of their responses has been to adjust their usual classroom pattern of single-day "lessons" and instead to set multiday goals for students on a regular basis. Instead of collecting computer-laboratory assignments each day, for example, they have found it more suitable to tell students that they must complete a given amount of material within a fixed period of time (two, three, or four days). The goals generally do not match the speed of the fastest students but do quicken the pace of the slowest workers. Students who are working at a slower pace take advantage of study halls and other free periods to obtain additional help from fellow students or from their teacher to complete their work.

Responsibilities and Challenges in Conducting and Planning Classroom Activities

If teachers are to incorporate the computer as an integral part of their mathematics classes, they will be faced not only with orchestrating lab experiences but also with planning and executing appropriately interfaced classroom lessons. One area of special concern is the nature of preparatory activities for the computer-lab explorations. Depending on the requirements of the lab activity, the teacher will include varying combinations of the following: an introduction to, or demonstration of, the software; a reminder on features of operating the hardware; and a discussion of applicable problem-solving strategies. Whatever teachers provide during the preparation period, however, collective student attention will be difficult to procure once the lab starts, leaving little opportunity to repair general misinformation or to provide additional instructions.

A second area of concern is that of conducting whole-class demonstrations on the computer. Mathematics teachers are unaccustomed to depending on electronic devices for their demonstrations, and they are sometimes reticent to center class discussion on computer output. Sometimes the computer output is unpredicted by the teacher, and besides, the use of the computer as a tool in the mathematics classroom requires new mathematical skills. For example, if the computer is to be used for calculations, graphing, or symbolic manipulations, how can one check the reasonableness of results? If the computer delivers eight-place decimals for its numerical calculations, how does one determine the needed degree of accuracy? Teachers who incorporate computer demonstrations into their lessons find that they can anticipate many of these difficulties if they regularly plan their lessons with the appropriate computer and software at hand.

In courses that focus on tool-assisted problem solving, a primary purpose of the classroom computer demonstration is to model a computer-based problem-solving process for students. When students have access to a variety of computer tools and representations (numerical, graphical, and symbolic), teachers not only must show students how to use each of the computer tools but also need to portray the flexible use of the tools and available representations. Their demonstrations should address the issues of the choice of appropriate representations, the selection of computer tools, and translations among representations through the use of several computer tools. As teachers conduct their computer demonstrations, they constantly face decisions concerning how much to explain about the problem-solving processes they were demonstrating and how much to leave to discovery. Although decisions like these are similar to those faced regularly by good mathematics teachers in traditional settings, the need to confront these decisions is accentuated in a setting that assumes easy student access to a powerful set of computing tools. Teachers can get cues on how much of the problem-solving process to specify in their demonstrations, however, through careful and reflective observation of students working on computer-laboratory explorations.

NEW ROLES AND RESPONSIBILITIES FOR STUDENTS

Even though the list of new roles and challenges for teachers in a computer-enriched mathematics classroom may seem formidable, no less a challenge lies in store for the students. Some of the new roles for students are directly related to engaging in group problem solving. If students are to work with partners in problem-solving tasks, they must learn to cooperate and collaborate. As the need arises for decisions about what programs to use, what strategies to employ, and what data to process, lab partners will need to negotiate a decision-making process. They will also learn to share the more routine tasks (keyboarding, checking the accuracy of input, checking the reasonableness of output). Leadership roles within a pair are likely to change as the keyboarder changes or as different software is used. The cooperative nature of work with partners can spill over into interactions with other pairs of students. We have regularly observed laboratory partners in the computer-based algebra course working in cooperative/competitive modes with other pairs, alternately offering technical assistance and engaging in good-natured races to be the first to discover a formula or produce an answer to the desired accuracy.

In the environment of cooperative problem solving, students can learn to use fellow students as legitimate resources for learning. This phenomenon is illustrated in the following field notes about the computer-based algebra course taken by a researcher who happened to be in a computer lab when the teacher was not present:

> Although the students in the lab had been told by their classroom teacher to work independently on the AREA AND PERIMETER lab, the students appeared to be unable to resist working with one another. Little by little, students began to call out their findings and check their work with one another. Working pairs and triples began to form as the students engaged rather enthusiastically in the task at hand. Some rather heated debates took place among close to half of the students in the lab toward the end of the period as students began to challenge the findings of one student who had been working alone.

Computer-laboratory explorations in pairs can also increase the need for students to communicate both orally and in writing. Oral communication is a necessity if joint decision-making is to evolve, and written communication is needed for computer-lab reports. These communication opportunities, rare in the traditional mathematics classroom, occur naturally in the computer-enriched problem-solving classroom.

Responsibilities and Challenges of Self-directed Learning

As students are exposed to computer-based problem-solving curricula, they will accept a greater amount of responsibility for their own learning. A first category of new learning skills deals with classroom and computer-lab behaviors. Although the questions of concern are reminiscent of those asked of a traditional curriculum, their answers have a different tenor. Students in computer-based problem-solving courses are apt to have the following concerns: "How do I complete open-ended tasks?" "How do I know if I'm correct?" "How do I take notes from a mathematics 'discussion'?" "How can I work effectively with a laboratory partner?"

Teachers can help students deal with their new responsibilities. They can provide frequent feedback on students' progress on open-ended tasks. They can assure students that different answers and techniques can all be correct. They can begin with giving specific directions on what notes to take during a mathematics discussion and gradually lessen the amount of direction they give. They can make specific suggestions to students on how to begin working with their partners, perhaps by specifying group structure (i.e., switch keyboarders every class), emphasizing the need for partners to check each other's work, and moving away from specific suggestions as students become more able partners in the group process. Teachers must expect the group process to evolve. They should resist the temptation to make decisions for the students. When confronted with a direct "How should I . . .?" question by one partner, they can direct the question to the other partner or ask, "Did you ask your partner?" They should encourage the use of other students as legitimate resources, asking, "Would you help that pair?" or "What did your neighbors get as an answer?" while conveying complete confidence that cooperation will follow.

A second category of new learning skills for students centers on their ability to assess how well they have learned. In traditional mathematics curricula, students feel they have learned the material if they can reproduce the procedures shown to them in class or if their answers match the ones given by the teacher (or in the back of the book). In order to prepare for tests, they redo problems they have previously completed. In tool-assisted problem-solving curricula, the primary goal is not the reproduction of modeled procedures. On tests and on assignments, students are often required to use the computer to apply their knowledge of strategies and mathematical representations to realistic situations. For many of the questions, there are many right answers. Students cannot practice the computer skills at home, since teachers are not often free to distribute the software. The techniques students have previously developed to assess their own learning are no longer as useful. They need to develop new methods for answering the questions "How do I know when I'm successful?" "How do I prepare for tests?" "What do I practice?" and "How do I interpret the teacher's feedback?"

Teachers can help students learn to assess their understanding in a tool-assisted problem-solving curriculum. They can teach students to check answers they have obtained graphically through an equivalent numerical or symbolic representation and to check computer results with periodic calculator computations. They can emphasize the need for ongoing active participation in laboratory explorations, for active reflection on problem-solving decisions, and for practice with what to do in unusual situations. Teachers can guide students in acquiring a new set of study skills (reading the text for understanding, reflecting on class notes, reviewing procedures used in completing computer-lab activities). When giving feedback, teachers will need to acclimatize students to the "phenomenon" of there being "many right answers," so that the students don't feel the need to reproduce their neighbors' results without understanding.

The changing role of the student in the computer-based problem-solving curriculum is no less complicated than the changing role of the teacher. If such curricula are to be used, teachers must provide students with guidance on how to act out their new roles.

CONCLUSION

Computer-enriched curricula can shift the focus of school algebra toward problem-solving and mathematical-modeling goals. This new conception of algebra begets a new conception of teaching and learning school mathematics. As students engage in exploratory computer-laboratory work, they will develop more robust views of the concepts, procedures, and strategies of mathematics. As their knowledge of the processes of mathematics expands, they will grow in their ability to work with, and learn from, peers as well as in their independence from the teacher. Teachers will face new challenges. They will function as facilitators in their students' development of facility with different computer tools, representations, and strategies. Before this ideal is attained, mathematics educators must collaborate in understanding the patterns of interaction between new conceptions of algebra and appropriate avenues for its teaching and learning.

REFERENCES

Fey, James, ed. *Algebra with Computers*. College Park, Md.: University of Maryland, 1986.

Heid, M. Kathleen, Charlene Sheets, Mary Ann Matras, and James Menasian. "Classroom and Computer Lab Interaction in a Computer-intensive Algebra Curriculum." Paper presented at the annual meeting of the American Educational Research Association, New Orleans, April 1988.

National Council of Teachers of Mathematics. *Curriculum and Evaluation Standards for School Mathematics*. Reston, Va.: The Council, 1989.

Sheets, Charlene, and M. Kathleen Heid. "Integrating Computers as Tools in Mathematics Curricula (Grades 9–13): Portraits of Group Interactions." In *Small Group Cooperative Learning in Mathematics: A Handbook for Teachers,* edited by Neil Davidson, pp. 265–94. Menlo Park, Calif.: Addison-Wesley Publishing Co., 1990.

The preparation of this article was supported in part by the National Science Foundation under NSF award numbers DPE 84-71173 and MDR 87-51499. Any opinions, findings, conclusions, or recommendations expressed herein are those of the authors and do not necessarily reflect the views of the National Science Foundation.

24

The Impact of Graphing Calculators on the Teaching and Learning of Mathematics

Gloria Barrett
John Goebel

THE presence of graphing calculators in high school mathematics classrooms will have a significant impact on the teaching and learning of secondary school mathematics in the 1990s. In NCTM's *Curriculum and Evaluation Standards for School Mathematics* (1989), it is stated that "calculators and computers with appropriate software transform the mathematics classroom into a laboratory . . . where students use technology to investigate" (p. 128) and "the teacher encourages experimentation" (p. 128). With a classroom set of calculators that graph functions and ordered pairs of data, solve equations, and graph and compute the parameters for regression lines, students will be able to investigate and explore mathematical concepts with keystrokes. Mathematics teachers will stimulate the process by presenting students with interesting problems and guiding them in their search for solutions. The graphing calculator will help foster what the *Curriculum and Evaluation Standards* describes as "the emergence of a new classroom dynamic in which teachers and students become natural partners in developing mathematical ideas and solving mathematical problems" (p. 128).

The use of the microcomputer has given some of us a head start on the "calculator revolution." For several years now, many teachers have used computers to enhance instruction. The microcomputer has not had the impact that many people predicted, however. There appear to be two primary reasons for this. First, many schools still do not have a computer in each mathematics classroom. Second, those teachers who have a computer to use in front of their classes have had trouble defining its role in the classroom. All too often the computer has not been promoted as a problem-solving tool. There is concern that students are not able to solve problems and discover relationships for themselves except where computers are available for their use, not just the teacher's use. Though some students have their own computers, software packages are expensive and are not interchangeable, so we cannot require students to do homework or take tests on computers.

Unlike computers, graphing calculators will soon be available to all students. Most students will be able to afford their own calculator. To ensure that all students are accommodated, schools could buy an entire classroom set of graphing calculators for the cost of one microcomputer. When available to every student, these calculators will change the way we teach many of the topics in the traditional secondary curriculum and enable us to focus more attention on introducing new topics and real-world applications into those courses.

Two areas of instruction in which the graphing calculator is likely to have the most immediate impact are solving equations and analyzing functions, topics that make up a major part of the present algebra and precalculus syllabi. An additional area in which these calculators can enhance instruction is data analysis. This topic is virtually nonexistent in the curriculum of most secondary schools, despite the recommendations of experts in documents such as NCTM's *Agenda for Action* and *Curriculum and Evaluation Standards*. With calculators to graph scatter plots and find equations of regression lines, students will be able to develop a much better appreciation for the applications of the functions they study as models of phenomena in their world.

SOLVING EQUATIONS

With the advent of the graphing calculator, we have available a greater variety of techniques for solving equations than previously. These additional techniques are more general than the traditional algebraic methods, since they are independent of the type of expression involved. Because the graphing techniques can be applied to any type of equation, students will be able to solve more interesting problems at every level in their study of mathematics. Teachers can develop lessons focused on problems that involve quadratic equations to use with students who may never study the quadratic formula. Presented in an algebra class, the same problem could be used to show students realistic applications *before* they learn computational techniques. Calculator utilities, therefore, will help teachers provide opportunities that develop mathematical power and move *all* students forward. It is anticipated that mathematics classrooms will become more exciting and that the problem-solving skills of all students will improve as they spend more time investigating and exploring to find answers to interesting problems.

Example 1. A travel agency advertises a special deal for group travel to the Super Bowl. Transportation is on a chartered plane that seats 200 passengers. If the group fills the plane, the price per ticket is \$150. The ticket price increases by \$1 for each empty seat. If the airline charges the travel agency \$18 000 for the plane, what is the minimum number of passengers needed for the agency to break even?

Solution. Students may need to discuss this problem in class in order to understand it completely. Group discussion helps students focus more clearly on what information they have and what they need to find. If necessary, teachers can guide the discussion by asking probing questions. Otherwise their role would be to serve as a moderator and become a partner in the search for a solution. Students often find it helpful to consider specific situations before they write an abstract mathematical formula. They may ask, for example, how much money the travel agency will receive if 190 passengers sign up for the trip. In answering this specific question, they will recognize that the answer is obtained by multiplying the number of tickets and the price of each ticket, in this case $190(150 + 10)$, since 190 passengers imply ten empty seats. In lower-level classes, students might choose to assign additional calculations to small groups and create a table of values to be used to answer the question. Using the calculator will facilitate this process. Other students will develop the expression for income, $(200 - x)(150 + x)$, where x represents the number of empty seats. They will then attempt to determine what value of x produces an income of \$18 000. That is, they will need to solve $(200 - x)(150 + x) = 18\,000$.

Students with no knowledge of the quadratic formula can easily solve this equation with a graphing calculator. One approach involves representing each member of the equation as a function. In this example we can let $f(x) = (200 - x)(150 + x)$ and $g(x) = 18\,000$. We can then graph these two functions simultaneously and determine the point of intersection. Students must determine the appropriate interval and specify the viewing window for their graphs. Clearly the number of empty seats must be a nonnegative number less than 200, the seating capacity of the plane.

When we graph $f(x)$ and $g(x)$ on the interval $[0,200]$, it is difficult to see exactly where the point of intersection occurs (fig. 24.1). Students can use the blinking pointer to determine that the solution to this equation is somewhere between 135 and 140 empty seats. (Students who have experimented with several specific values may have a reasonable approximation for the solution and decide to graph over a smaller domain initially, as in fig. 24.2.)

To determine a more exact value, students can redraw the graphs over a smaller domain, or they can use a zoom feature of the calculator. Practically, since x must be a whole number, they need only to trap the point of intersection between the two integers 137 and 138. Since the students know

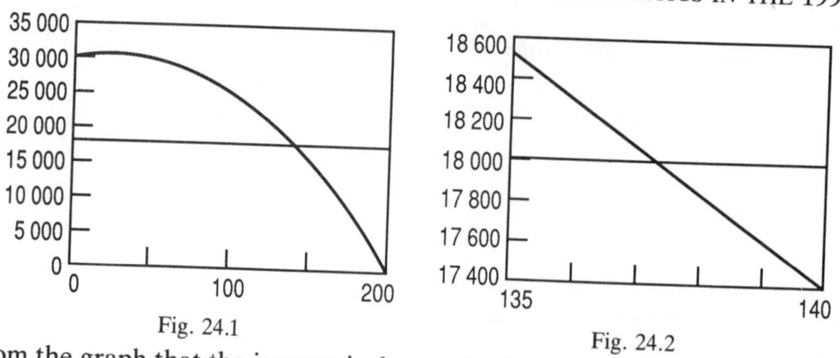

Fig. 24.1

Fig. 24.2

from the graph that the income is decreasing for these values of x, it is clear that 138 empty seats will not produce the needed income of \$18 000. The travel agent must have 137 or fewer empty seats, so the minimum number of passengers needed to break even is $(200 - 137)$, or 63.

Although the equation above could be solved algebraically with the use of the quadratic formula, the graphing approach enables students to understand more clearly the connections between algebraic equation solving and graphical representations of the algebra. In this example, students who have examined the graph understand why 138 empty seats is too many, since they have actually seen that 138 produces an income value below what is required. The graphing approach also enables students to do much more than answer one specific question. It enables them to explore the answers to "what if . . .?" questions that might arise from this problem. For example, how many tickets must be sold if the break-even amount is \$16 000 or \$20 000? How many passengers produce the maximum income to the travel agency? What happens if the ticket price increases by \$2 for each empty seat?

An alternative approach for solving equations of the form $f(x) = g(x)$ is to define a new function $h(x)$ to be $f(x) - g(x)$, use the calculator to graph $h(x)$, and find the value of x for which $h(x) = 0$. In this example this is equivalent to setting up a profit function $p(x) = (200 - x)(150 + x) - 18\,000 = f(x) - g(x)$. Students could graph this function, determine the value of x that makes $p(x) = 0$, and then analyze what integer value of x is the correct answer to the problem.

Calculator methods for solving equations are very quick and accurate, but they will disturb those of us who believe that every student should be able to solve equations by hand. With access to new technology, this belief is becoming very difficult to justify and support. If repeated solving of equations and inequalities added to students' understanding of mathematics as a way to represent reality, or added more to students' understanding of mathematical systems, then the time spent solving those equations by hand could be justified. If not, a major part of our present curriculum is geared toward a skill that a calculator can perform quicker and more efficiently.

ANALYZING FUNCTIONS

Obviously the graphing approach for solving equations is very powerful. Students with access to computers and graphing software may already be using these techniques; in the 1990s every student will have the power to do the same thing with a graphing calculator. Consequently, writing functional expressions and interpreting graphs will become major topics in the calculator-enriched classroom. It is worth noting that an awareness of domain and range is even more important for graphical solutions than for traditional algebraic methods. Students will use both their understanding of the problem and their knowledge of general characteristics of elementary functions to specify appropriate viewing windows for the functions they graph.

Though graphing is a major topic of study for most college-intending students, the focus is seldom on using graphs to model real-world phenomena. The graphing calculator enables us to draw very accurate graphs so that we can incorporate more interesting problems in our courses. For example, optimization problems that require finding extreme values of a function can be included in a high school algebra or precalculus course. To solve such problems, students can graph the function, zoom in on extreme points, and use the trace feature to find very accurate approximations for extreme values.

It seems clear that our approach to analyzing functions will change. Whereas we once tried to find important points, including the zeros and extreme points, in order to graph, we will now use the calculator first to graph the function and then to analyze the important features of the graph. By specifying different dimensions of the viewing window, we can observe additional characteristics of a function, including its global and local behavior, limits, and asymptotes. These graphs can also be used to enhance understanding when we teach more traditional topics. With calculator graphs, students can explore the cofunction relations in trigonometry, the rate of growth of an exponential function, the relationship between the graphs of a function and its inverse, or the behavior of a rational function at x-values close to those that make the denominator zero. Teachers, of course, will need to guide the exploration, but students will have the opportunity to discover for themselves many of the relationships that we want them to learn. Furthermore, this learning will come from understanding rather than from memorizing.

DATA ANALYSIS

One of the most powerful applications of functions is to provide empirical models for the relationship between two variables. Students who learn techniques of data analysis can perform experiments, collect data, and use the calculator to help analyze their data. The graphing calculator allows us to enter a set of ordered pairs of numbers and then plot them on the graphing screen. If students decide after viewing the scatter plot that a linear model is appropriate, the calculator will determine the equation of the regression line for the data and graph the regression line with the ordered pairs.

When an examination of a scatter plot reveals that the relationship between two variables is not linear, students must draw on their knowledge of various functions and their properties to conjecture what function would provide the best model for the data. The calculator can then be used to reexpress one or both of the coordinates in a way that produces linearity. A knowledge of function composition and inverses will guide the reexpression process. After students succeed in linearizing data, they can use the calculator to fit a regression line through the reexpressed data and derive a model for the observed relationship. This process, which often involves much exploration by students and seldom results in a single correct answer, is illustrated by the following example.

Example 2. The data in the following table show the population per square mile in the United States from 1790 to 1980. Find a mathematical model (that is, a functional expression) that can be used to predict the population density in the year 2000.

Year	1790	1800	1810	1820	1830	1840	1850	1860	1870	1880
People per Square Mile	4.5	6.1	4.3	5.5	7.4	9.8	7.9	10.6	10.9	14.2

Year	1890	1900	1910	1920	1930	1940	1950	1960	1970	1980
People per Square Mile	17.8	21.5	26.0	29.9	34.7	37.2	42.6	50.6	57.5	64.0

Solution. We shall let x represent the years since 1700 and y the population per square mile. When students use the calculator to plot these ordered pairs, they notice that the scatter plot is not linear (fig. 24.3). From the curvature in the scatter plot and their knowledge of functions, they might speculate that the relationship could be modeled by either an exponential or a power function. Students who have some knowledge of the growth phenomenon might argue that the exponential model is more appropriate. Proceeding under this assumption, they would then apply their knowledge of inverse functions and attempt to straighten the curvature by taking logarithms of the y-values. (If no one makes a strong case for the exponential function, students might choose first to perform log-log reexpression. Teachers should encourage experimentation and allow false starts.) The calculator will compute the natural logarithms as ordered pairs are entered, for example $(90, \ln 4.5)$. These points can then be graphed to determine whether the reexpression successfully linearizes the original curvature. Since the semilog reexpression does result in a linear scatter plot (fig. 24.4), the calculator can now be used to find the equation of the regression line through the points (x, ln y).

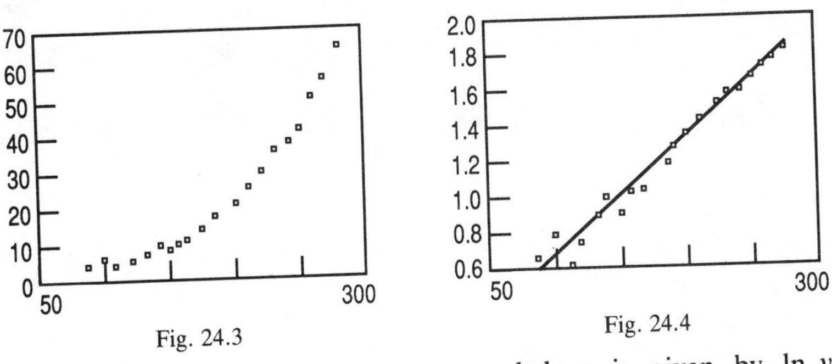

Fig. 24.3

Fig. 24.4

The linear relationship between x and ln y is given by $\ln y = 0.015x + 0.049$. A functional expression for y in terms of x is $y = 1.050e^{0.015x}$. Assuming that this relationship holds, students can now use this function to predict the population density in the year 2000. Substituting 300 for x in the model, we predict a population density of 94.5 people per square mile.

Analyzing data to find mathematical models for relationships between variables and to make predictions provides almost unlimited opportunity for students to apply the mathematics they learn to meaningful problems. In the preceding example, students use a network of mathematical knowledge, including exponential functions, inverse functions, and logarithms, to determine the model they need to predict future population density. Furthermore, problems such as this one provide special opportunities for teachers to encourage group discussion of open-ended questions. Students should be able to support their choice of an exponential model, explain why they expect semilog reexpression to linearize the data, and discuss the assumptions they make when they use this model to predict.

CONCLUSION

There is no doubt that graphing calculators will bring about major changes in the way we teach secondary school mathematics. As these calculators become accessible to all students, we will be able to make changes in our classrooms that broaden and enrich our students' experiences in mathematics. It is our responsibility to design lessons that will make mathematics more interesting and stimulating for students. As algebraic manipulation gives way to new content, exploration, and discovery, students will become more enthusiastic learners. Not only will they better understand mathematics, but they will also discover their power to use mathematics to solve problems that influence their lives.

25

Enhancing Mathematics Teaching and Learning through Technology

Franklin Demana
Bert K. Waits

THIS article illustrates how technology can enhance the teaching and learning of mathematics in the 1990s in a manner consistent with the NCTM's *Curriculum and Evaluation Standards for School Mathematics* (National Council of Teachers of Mathematics 1989). The greatest benefits seem to come from interactive technology that (1) is under student and teacher control, (2) promotes student exploration, and (3) enables generalization. Computer graphing utilities make graphing a fast and effective problem-solving strategy. With this power we can now do the following:

- Graph numerous functions quickly and establish common properties of classes of functions, have students explore and discover mathematical concepts, and use graphs to solve problems. (In the past these activities were usually not possible because of the time required to produce accurate graphs.)
- Use realistic applications involving complicated algebraic representations. (In the past applications were contrived and restricted by the limitation of the algebraic techniques available in the curriculum.)

Classroom instructional models that encourage students to be active partners in the learning process are a natural consequence of this approach. Technology available today should dramatically change the way that mathematics is both *taught and learned*. For example, the study of domain, range, inverses, geometric transformations, solutions to equations, inequalities, systems of equations and inequalities, and applications can be accomplished more effectively with a technological approach. Powerful geometric representations of problem situations can be easily added to the usual algebraic representations. Thus, the power of visualization can be used to study mathematical concepts and ideas.

The technology-based approach to the teaching and learning of mathematics described in this article was piloted and field-tested in the Calculator and Computer Precalculus (C^2PC) Project, funded in part by the National Science Foundation, British Petroleum (Ohio), the Ohio Board of Regents, and Ohio State University. C^2PC teachers use two important technologically driven instructional strategies. Students in classrooms with a single computer participate in interactive lecture-demonstration. Computer laboratories and classrooms where students have graphing calculators provide a setting for a guided-discovery instructional model. Teachers ask appropriate questions and provide supporting activities to help students understand or discover important mathematical concepts. Graphing calculators are used regularly on homework assignments. C^2PC students develop strong intuition and understanding about functions, the foundation that helps make the study of calculus, advanced mathematics, and science successful.

OPPORTUNITIES AND CHALLENGES OF A TECHNOLOGICAL APPROACH

In the past careful numerical and analytical techniques were used to produce accurate graphs that were rarely used. Now accurate graphs are obtained quickly with technology and are used to study the numerical and analytical properties of a function. Students are more motivated to ask and answer questions about properties of a function that are generated by viewing its graph. This questioning helps ensure that computer-drawn graphs are correct.

Mathematical Exploration

The first example illustrates how a single classroom computer can make the lecture-demonstration approach highly interactive and visual.

Example 1. Draw the graphs of $y = x^3$, $y = 5 + x^3$, $y = 5 + (x - 6)^3$, and $y = 5 + (x + 3)^3$ in the same coordinate system. Predict the graph of $y = v + (x - h)^3$ for real numbers v and h.

Solution. The four computer-drawn graphs are displayed in figure 25.1. Teachers can use "What if . . .?" questioning to help students understand the role of v and h in $y = v + (x - h)^3$.

The speed and power of computers and graphing calculators can help students develop a geometric interpretation of the role of a, b, c, and d in producing the graph of $y = af(bx + c) + d$ from the graph of $y = f(x)$. Functions like $y = 1/x$, $y = \sqrt{x}$, or $y = \sin x$ become basic blocks from which complicated functions are built and understood. After several investigations of this type, students can be expected to use their knowledge of a

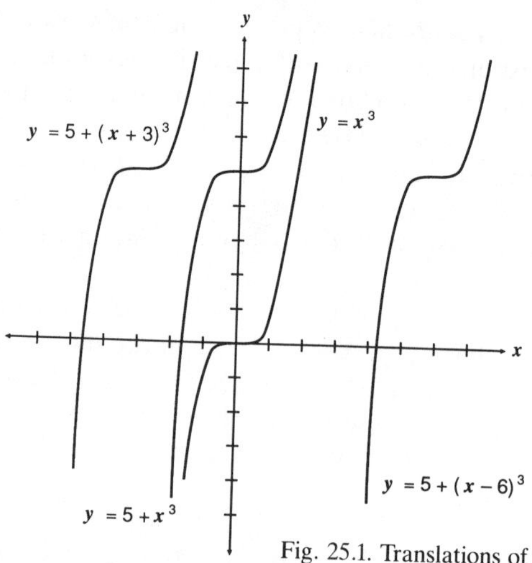

Fig. 25.1. Translations of the graph of $y = x^3$

few basic graphs and the geometric insight gained in the classroom with a computer to sketch graphs of complicated functions *without* using a graphing utility (such as either computer graphing software or a graphing calculator).

A guided-discovery instructional strategy is possible in a computer laboratory with up to three students at each computer, or in a classroom with every student using a graphing calculator. For example, with a polar graphing utility students can easily obtain the graphs of $r = m \sin(nt)$ for various values of m and n and then deduce the effect of each factor. After a little exploring students conjecture that the graph of $r = m \sin(nt)$ is an n-leaf rose for odd integers n and a $2n$-leaf rose for even integers n and that the lengths of the "rose petals" are directly related to m (fig. 25.2). Determining the exact length of the "rose petals" using trigonometry and investigating the graphs when n is replaced by a *rational* number are interesting student exercises. For example, what is the graph of $r = 5 \sin 2.5t$?

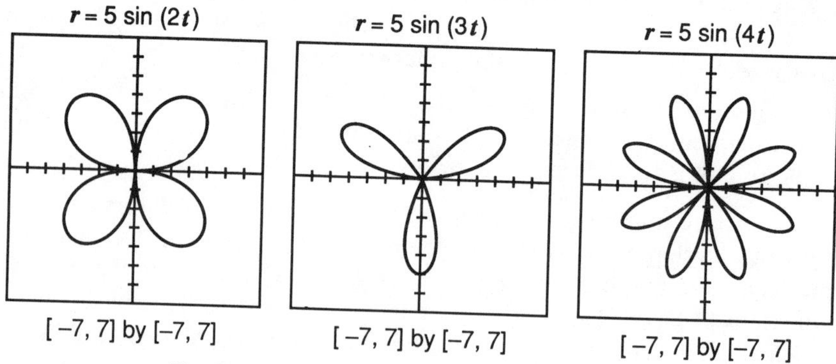

Fig. 25.2. Polar graphs of four-, three-, and eight-leaf roses

Teachers using a guided-discovery approach must carefully select a sequence of visual experiences that will help students understand or discover a given mathematical concept or idea. Creating a lesson plan with many pictures and as few words as possible is an effective way to prepare for a guided-discovery lesson.

Shift in Instructional Emphases in Light of Technology

Technology not only affects how students learn but also raises expectations of new and different types of understanding on the part of students. In particular, in our work with graphing calculators in advanced algebra and precalculus classes, student understanding of the effects of scaling on graphs, the interplay between mathematics and its application, error analysis, and multiple representations took on increased importance.

Understanding the Effects of Scaling

Scaling is a crucial issue. Graphing utilities allow students to specify a viewing window or rectangle when drawing graphs. A *viewing rectangle* $[a, b]$ *by* $[c, d]$ is the rectangular portion of the coordinate plane determined by $a \leq x \leq b$ and $c \leq y \leq d$.

Students and teachers must adapt to the fact that the shape of a graph depends on the viewing rectangle in which it is viewed. For example, figure 25.3 illustrates that the graph of $y = x^2$ can appear to be very flat or very steep depending on the viewing rectangle used. Notice that the horizontal dimension $b - a$ is five times the vertical dimension $d - c$ in the second viewing rectangle of figure 25.3. Such viewing rectangles provide a poor view of the graph of $y = x^2$ but provide instructive views at local extrema of functions where function values change slowly compared to independent-variable values. Similarly, if function values are increasing rapidly, viewing rectangles whose vertical dimension is greater than their horizontal dimen-

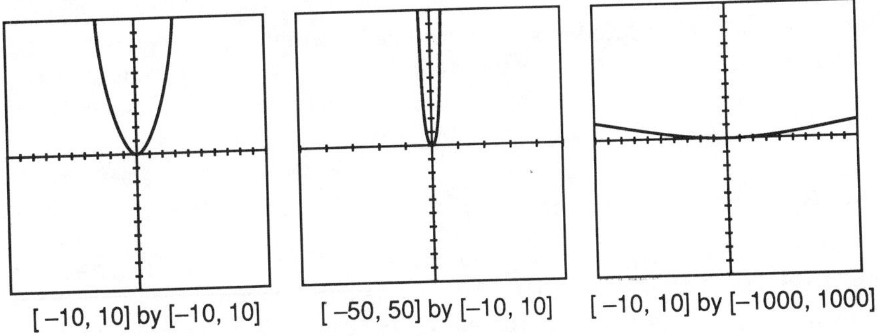

Fig. 25.3. Three different views of the graph of $y = x^2$

sion, like the third graph of figure 25.3, give good views. Such viewing rectangles give good views near vertical asymptotes of functions.

Students and teachers must learn how to choose viewing rectangles that give good pictures as they explore using powerful graphing-utility features, one of which is called *zoom-in*. This technique places particular importance on scaling. Zoom-in permits a portion of a graph to be magnified and analyzed. Once a graph is on the display screen, zoom-in allows the user to quickly solve related equations, inequalities, and relative maximum and minimum problems and to "see" that locally a continuous function is a straight line.

Zoom-out is a way to quickly determine the global or asymptotic behavior of a graph by viewing it in several large viewing rectangles, each containing the previous one (Demana and Waits 1990). The zoom-in and zoom-out features help give a "complete" picture of the behavior of graphs of relations. A graph that shows all the relevant behavior is called a complete graph. Sometimes several graphs are needed to display the complete behavior of relations.

This technique can be an extremely powerful and general method to solve equations and determine maximum and minimum values of functions without using the tools of calculus. Zoom-in is a fast graphical refinement of numerical approximation techniques such as the bisection method. These geometric approximation techniques can be used effectively to foreshadow the study of calculus and provide a firm intuitive foundation for a rigorous study of advanced mathematics.

Zoom-in creates a sequence of nested viewing rectangles that "converge" on a desired point. Each new viewing rectangle is contained in the previous viewing rectangle. This is accomplished by simply changing the parameters of the viewing rectangle, that is, the scaling of the rectangles.

The ability to obtain a graph of $y = f(x)$ quickly makes it very natural to solve the equation $f(x) = 0$ by zooming in on the corresponding x-intercept. Zoom-in provides a way to solve equations that students could not previously solve with only algebraic techniques, as illustrated in the next example. Solving inequalities soon becomes a geometric problem of finding where one graph is above or below another, or where one graph is above or below the x-axis. Such a technique requires the student to understand how to scale the portion of the graph to be viewed.

Considering realistic applications

Technology makes realistic problems accessible to students much earlier in their study of mathematics and removes the barriers to high-quality problem-solving activity that are caused by a lack of familiarity or facility with algebraic techniques. The next example is an important real-world problem that in the past was usually reserved for calculus students. However, this problem is now very accessible to algebra students using computer graphing.

Example 2. The Benders can afford to pay \$600 a month for a 25-year home loan. What annual percentage rate (APR) will permit them to purchase a \$65 000 home?

Solution. Let x be the *monthly* interest rate; then $12x$ is the APR. It is not difficult to establish that x is given by

$$65\,000 = 600\,\frac{1 - (1 + x)^{-300}}{x}$$

(Waits 1979). Since there is no exact solution to this equation, solving it requires a numerical method. A graphing-based method is very natural. Let $f(x)$ be the left-hand side and $g(x)$ the right-hand side of the equation above. To solve $f(x) = g(x)$ graphically means to graph $y = g(x)$ and $y = f(x)$ in the same viewing rectangle and then determine the points of intersection. It is difficult to find a viewing rectangle that gives a good view. This is an example where viewing rectangles with larger vertical dimensions are needed. Graphing can be restricted to the first quadrant because the x values represent possible monthly interest rates and the y values represent possible dollar amounts for the loan. Thus, the maximum y value for a viewing rectangle must be greater than 65 000, and it is reasonable to assume x is less than 0.1 (10 percent a month). The first view of figure 25.4 shows the graphs of

$$f(x) = 65\,000 \text{ and } g(x) = 600\,\frac{1 - (1 + x)^{-300}}{x}$$

in the [0, 0.1] by [0, 100 000] viewing rectangle and suggests that 0.01, or 1%, is the only solution. The middle view of figure 25.4 is a magnification

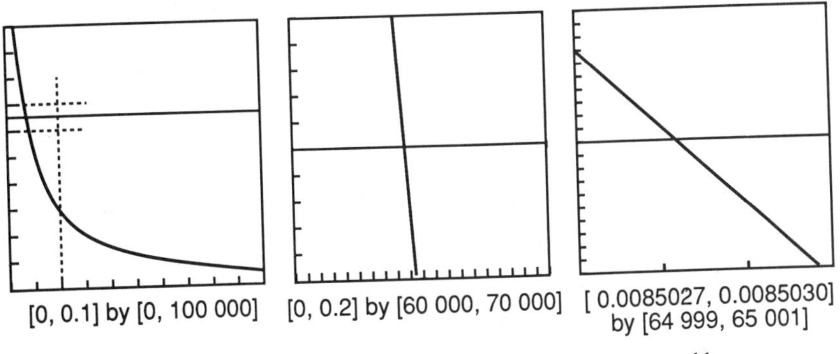

Fig. 25.4. Three geometric representations of the home-loan problem

of the region within the small rectangle in the first view. The last view is the result of several graphing zoom-in steps and can be used to determine that the monthly interest rate x is about 0.0085028 with error less than 0.0000001, the distance between the horizontal scale marks. Thus, the desired APR of the home loan must be approximately 10.20 percent.

The role of error analysis

Notice that the vertical dimension of the viewing rectangle of the last view of figure 25.4 is about 6.66×10^6 times the horizontal dimension. Thus, changes in y that appear to be small in this view are actually quite large when compared to the x-scale. For this reason, the error in y (the loan amount) is much larger than the error in x (the monthly interest rate).

Students applying technology to solve problems need to understand and control the error in their solution. Using graphing zoom-in to solve equations provides an excellent geometric vehicle for discussing error. Students' ability to judge the reasonableness of numerical solutions found using technology needs to be increased. Emphasis on exact solutions should be reduced, since approximate solution methods deserve equal, if not more, emphasis. Numerical approximation methods and associated error analysis become more meaningful to students in a geometric setting.

Promoting multiple representations

Notice that the function g decreases in the viewing rectangle of the first view of figure 25.4. It is a good exercise to have students observe and interpret this behavior in the problem situation. Discussing how the loan amount decreases as the interest rate increases helps make the connection between the algebraic and geometric models for this problem situation.

The ability of students to operate within *and* between different representations of the same concept or problem setting is fundamental in effectively applying technology to enhance mathematics learning. For example, the important connections among a point (a,b) on a graph, an associated functional representation $b = f(a)$, and interpreting a and b in a related problem situation are not usually well established using conventional methods. More attention needs to be given to the topic of translating between algebraic and geometric representations of problem situations.

TRANSLATING BETWEEN ALGEBRAIC AND GEOMETRIC REPRESENTATIONS

Graphing utilities make the addition of a geometric representation to the usual numerical and algebraic representations very natural. The availability of a geometric representation gives students and teachers the opportunity to explore and exploit the connections between algebraic and geometric representations. The interplay between algebraic and geometric representations is illustrated in the following maximization problem.

Example 3. Suppose squares of side length x are removed from a 25-inch-by-40-inch piece of cardboard, and a box is formed as shown in figure 25.5. Use a complete graph of the algebraic model V to determine what value of x maximizes the volume.

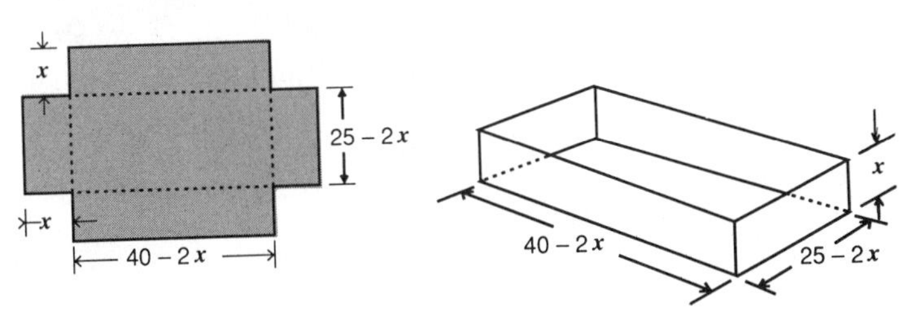

Fig. 25.5. The volume of a box problem

Solution. It is easy to establish that $V(x) = x(25 - 2x)(40 - 2x)$ is an algebraic representation of the volume as a function of x. A complete graph of $y = V(x) = x(25 - 2x)(40 - 2x)$ is shown in figure 25.6. Given that x must be positive and that $2x$ must be less than 25 (the side length of the cardboard), the values of x that make sense in this problem situation are $0 < x < 12.5$, and only the first-quadrant portion of the first graph of figure 25.6 that is above the x-axis with $x < 12.5$ represents the problem situation. Therefore, the second graph of figure 25.6 is a geometric representation of the problem situation. Figure 25.6 strongly suggests that there is a maximum value of V of about 2300, and it occurs when x is about 5. Zoom-in can now be used to determine the solution to a very high degree of accuracy.

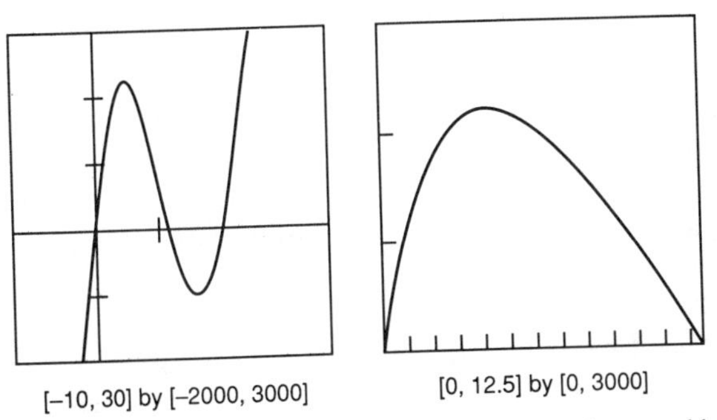

Fig. 25.6. Geometric representations of the box-volume problem

We find that considerable discussion is necessary for students to readily associate the coordinates of the "maximum point" with a solution to this real-world "maximization" problem. The connection between the coordinate representation of points (a,b) of the graph of V and $b = V(a)$ must be established. That is, in (a,b), a represents a possible side length of a removed square and b represents the corresponding volume of the resulting box only for certain values of a and b. Such discussion helps establish the connections among the graphical representation, the algebraic representation $y = V(x)$, and the problem situation. Now, if (a,b) are the coordinates of the highest point, students can see that the maximum volume is $b = V(a)$ and a is the side length of the removed square. Such connections must be carefully developed with many examples during the school year. Once this kind of activity is well established, it is easy to move to zoom-in as a procedure for determining very accurate solutions to these types of problems.

SOME LESSONS LEARNED

Many C^2PC students seem almost happy to go through algebraic processes to confirm conjectures discovered visually through computer graphing. For example, when asked to sketch a complete graph of

$$f(x) = \frac{x^3 - 13x^2 + 38x - 24}{x - 1}$$

without technology, students proceed as follows. First, polynomial division is used to write

$$f(x) = x^2 - 12x + 26 + \frac{2}{x - 1}.$$

Then, a graph of $y = x^2 - 12x + 26$ is sketched and a portion near $x = 1$ is erased (fig. 25.7). Next, the graph of $y = \dfrac{2}{x - 1}$ is sketched and the portion away from $x = 1$ is erased. Finally, students blend these two pieces to obtain a very good rough sketch. They then check the sketch by using a graphing utility.

A great deal of algebraic "drill" can be disguised, and therefore practiced, in interesting computer-graphing-generated activities. Students seem to be receptive to algebraic activities and other important basic-skill practice when they are not the focus of a lesson. This is a frequent outcome of the use of technology in the mathematics classroom.

On the basis of our experiences, we believe there is great benefit to students when a "revisitation" approach to problems is used across courses as well as within courses. Learning is more efficient, stable, and permanent

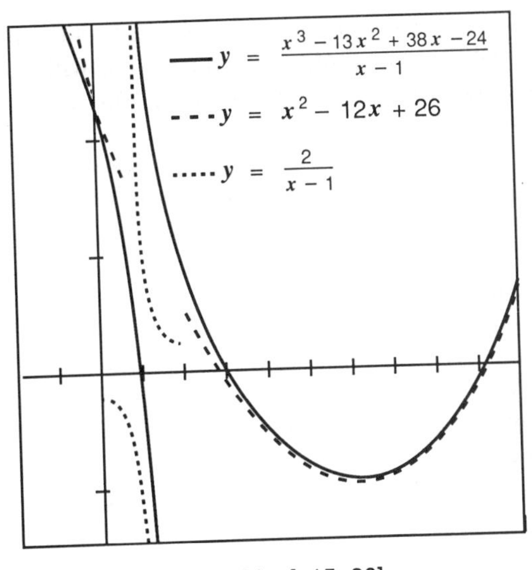

[–1, 10] by [–15, 30]

Fig. 25.7. A decomposition of f

because it is connected with what is already known. For example, the volume problem of Example 3 can first be explored in middle school mathematics classes by measuring and creating real paper models. Prealgebra and algebra students can return to the problem numerically and use tables to study the problem and establish the concept of variable (Comstock and Demana 1987; Demana and Leitzel 1988). Advanced algebra students can return to the problem and obtain a geometric representation of both the problem situation and the algebraic representation with computer graphing. Precalculus students can then use zoom-in to solve the maximization problem geometrically. Finally, when students see this problem again in calculus, they will have a rich, intuitive, geometric basis on which to apply the rigorous, formal techniques of calculus.

SUMMARY

The increasing use of technology in business, industry, and science necessitates changes in the public's expectations of school mathematics. More attention needs to be given to issues such as scaling, numerical analysis, error, and translating across representations in teaching. The easy availability of technology-generated models makes multiple representations of problems practical. Analyzing a problem situation through both algebraic and geometric representations deepens students' understanding about the problem. Instead of the usual negative attitude about word problems, students gain more confidence about problems with the added technique of analyzing and solving them graphically. Word problems seem less mysterious and not as formidable with the addition of a geometric representation and powerful graphical problem-solving methods.

Deeper understanding and an intuition about functions are important by-products of a technologically rich approach to the teaching and learning of mathematics. Solution methods based on computer graphing open the door to important realistic problems that will serve students well, both in postsecondary study and in the work place. The power of visualization helps students question, conjecture, and discover important mathematical concepts.

The use of technology gives rise to a much richer classroom environment and fosters student exploration and investigation. Exciting new teaching approaches are possible with the use of technology. These approaches make teachers and students active partners in a rewarding, enjoyable, and intensive educational experience.

REFERENCES

Comstock, Margaret, and Franklin Demana. "The Calculator Is a Problem-solving Concept Developer." *Arithmetic Teacher* 34 (February 1987): 48–51.

Demana, Franklin, and Joan Leitzel. "Establishing Fundamental Concepts through Numerical Problem Solving." In *The Ideas of Algebra, K–12,* 1988 Yearbook of the National Council of Teachers of Mathematics, edited by Arthur F. Coxford, pp. 61–68. Reston, Va.: The Council, 1988.

Demana, Franklin, and Bert K. Waits. *Precalculus Mathematics—a Graphing Approach.* Reading, Mass.: Addison-Wesley Publishing Co., 1990.

National Council of Teachers of Mathematics. *Curriculum and Evaluation Standards for School Mathematics.* Reston, Va.: The Council, 1989.

Waits, Bert K. "The Mathematics of Finance Revisited through the Hand Calculator." In *Applications of Mathematics,* 1979 Yearbook of the National Council of Teachers of Mathematics, edited by Sidney Sharron, pp. 109–24. Reston, Va.: The Council, 1979.

26

Teaching Mathematics with a Vision: Integrating Computers into Instruction

Marc Swadener
William Blubaugh

RESEARCH has indicated that "recontextualizing the curriculum" (Cole and Griffin 1987) can bring many mathematical tasks within the reach of *most* students—tasks that previously were considered beyond their capabilities. Over the past twenty years investigations into the lack of progress in raising the overall achievement in mathematics have found that the *context* in which mathematics is taught is as important as what is taught (Cole and Griffin 1987). Cole and Griffin state that learning mathematics in isolation is counterproductive. Changes in the context of mathematical learning can increase achievement more than dealing solely with what is taught, as has been done in the past. This change in context means not merely dealing with the physical environment of the classroom. The *intellectual context* in which mathematics is considered *must* be broadened in order to enhance learning and retention. Recent advances in computer, communications, and video technology—and the increasingly blurred distinction between them—promises that technology can play a significant role in recontextualizing schooling, especially with respect to the teaching and learning of mathematics. Technology, which is broadly used outside education, offers increasingly versatile tools that can supplement (not replace) the capabilities of the learner (be the learners teachers or students).

As an example of a new context for mathematics instruction and the role that technology can play in it, consider the topic "world food distribution and consumption." After collecting data, students could construct concrete representations of the numbers associated with population size and food consumption by using a computer simulation in which the dots on the paper could represent the population of a "community."

One way of doing this is to use a word processor to fill a page with colons (:). Consider the number of dots on the page (two dots per colon). Discuss the "best" way to count the dots. Since there is a regular pattern of dots on the paper, suggest that a systematic (mathematical) procedure could be used to count the dots. A spreadsheet could be used to develop relationships between factors associated with the size of the paper, the size of the characters (points or pitch), the number of lines, the size of the margins, and other "variables" associated with the simulation. Properly constructed, the spreadsheet automatically performs calculations based on these variables (as defined by the learners) and instantaneously updates the calculations when new values of the variables are entered. Learners will note that some quantities in the calculations are not fixed. That is, the number of characters on a page is *dependent* on the number of dots per character, characters per centimeter, lines per centimeter, and the size of the paper and margins. Concepts such as *independent variable, dependent variable, formula, algebraic expression, operations, order of operations,* and *degrees of freedom* will become clearer to the students. Producing the template is a dynamic process that generates considerable discussion and problem-solving activity.

The finished spreadsheet template given in figure 26.1 applies to this exercise. In this template students determine and enter formulas into cells B11, B12, B14, B15, B17, B18, B20, and B21 (cells are identified by column [letter] and row [number]). These cells represent *dependent* variables and are in **bold** print. In other cells (B3, B4, B6, B7, B8, B9, B10, B13, and B16) students enter actual numerical values. These *independent* variables (cells) are in normal print. In column A, students enter the descriptions of the quantities in column B. To use the fully developed template students need only enter values for the independent variables. After each entry the computer recalculates all dependent variables.

Students should be reminded that the reason for building the template is to provide a *concrete* representation of population size in the *context* of "world food distribution." Lines 20 and 21 are added to the template (see figure 26.1) to accomplish this representation.

Many additional questions can be asked:

1. How many head of cattle would be required if, for one year, each person in the United States ate one "quarter pounder" a week?
2. How much feed would be needed to raise the necessary cattle?
3. How much fertilizer would be used? What other energy sources would be used in producing the needed beef?
4. What is the efficiency of this method of nourishment? (*Efficiency* is a purposely vague term. Students are to develop definitions of efficiency and proceed with analyses accordingly.)

To answer these questions, additional lines could be added to the existing template (or a new template could be constructed). Added lines could

	A	B	
1	Item	Quantity	Explanation of entry in column B
2	Size of paper:		
3	Width (cm)	x	student entered value
4	Length (cm)	x	student entered value
5	Margins:		
6	Top (cm)	x	student entered value
7	Bottom (cm)	x	student entered value
8	Left (cm)	x	student entered value
9	Right (cm)	x	student entered value
10	Number of lines per cm	x	student entered value
11	Length of usable paper (cm)	**a**	B4 – B6 – B7
12	Number of lines per page	**b**	B10 · B11
13	Number of characters per cm	x	student entered value
14	Width of usable paper (cm)	**c**	B3 – B8 – B9
15	Number of characters per line	**d**	B13 · B14
16	Number of dots per character	x	student entered value
17	Number of dots per line	**e**	B15 · B16
18	Number of dots per page	**f**	B12 · B17
19			
20	How many pages to represent 1 million?	**g**	1 000 000 / B18
21	How many sheets to represent U.S. population?	**h**	265 / B20
22			

Fig. 26.1. Spreadsheet template

include the amount of beef consumed by each person each week, the number of weeks (fixed at 52, or variable), the population considered, the amount of usable beef produced from each head of cattle, or the "quality" of the beef produced. Additional lines could provide values for calculations based on student developed formulas and could yield the necessary results to answer the questions under investigation.

Further analyses could focus on the efficiency of food production (producing meat versus grain), the "quality" of the food produced (nutrition), marketing, distribution systems, and wastes involved in production and consumption. The students could develop a generalized template for computing "calories in/calories out" efficiency. A more global analysis could be considered by obtaining information about lifestyles and nutritional patterns in different cultures. This information could be obtained through a computer based information system and result in an analysis of the "interdependence of communities." At each stage of the analysis the original spreadsheet template could be either augmented or a new one constructed to "get a handle on" the magnitude of the numbers involved. Additional computer capabilities (graphing, statistics, graphics, and mapping) could be introduced. The computer would do the calculations and allow students to deal with "What if . . .?" questions. This exercise could result in an introspective look at the food consumption in the school and the resulting waste production. Students might develop a proposal to the school council for a long-range plan to deal with solid wastes and for the development of a school recycling center. They would be studying mathematics by attempting to conceptualize large numbers.

This example illustrates how the use of a computer spreadsheet could assist in reconceptualizing the term *mathematical problem*. In addition to spreadsheets, other computer packages, such as symbolic manipulators, graphing software, programming languages (Logo and others), data-manipulation software (including statistical and database), geometric and algebraic exploration software, and expert systems, have great potential for actively involving learners in mathematics teaching and learning. Clearly the creative use of symbolic manipulation software can assist in creating a new context for teaching mathematics. In a very real sense, the ability to use computer technology will become as much of a basic mathematical skill as memorizing "basic facts" was in the past.

Many current students lack investigation and exploration skills. The capability of computers to assist student discovery and conjecture is obvious. Computers provide instant calculations and rapidly generate graphics with which students can make and test conjectures. Many mathematical software packages are open-ended tools, adaptable to a range of learning and teaching needs and objectives. The degree of formal instruction versus investigation and conjecturing can be regulated by the teacher.

Software is available that will solve most of the exercises in today's mathematics textbooks. The widespread availability and use of mathematical-manipulation software will result in significant changes in the emphasis and paradigms used in school mathematics. Bollinger (1986) concluded that 75 percent of all problems in high school algebra could be solved completely or partially by symbolic manipulation software. Studies have shown that students learn more mathematics in less time with broader conceptual understanding using symbolic-manipulation software than from traditional instruction (Palmiter 1986). Future mathematics instruction should focus on concepts and applications. Proper use of symbolic manipulation software with application problems will change the focus of instruction and assist students through a conceptual and applied understanding of real-world mathematics.

Any contexts in which numbers are inherent are appropriate for making mathematical conjectures; this includes supply and demand, design, transportation, finance, consumer issues, economic development, pollution, construction, marketing, and so on. All such situations include numerous conditions for both mathematical and nonmathematical conjectures. One key in recontextualizing the mathematics curricula is examining the teacher's role in teaching and learning. The teacher's role must be to promote continuous active involvement of learners in "doing" mathematics.

What specifically needs to be done? First, mathematics teachers must be learners along with their students. Teaching and learning mathematics is not (and never was) an observer sport. The more learners become involved in mathematics, the more they will learn. Second, teachers need an inquiring mind and a vision of the future of mathematics and mathematics instruction. Third, teachers must feel comfortable using computer technology and have an awareness of applications and how computers can be effectively integrated into learning situations. The use of computer technology in mathematics teaching and learning should be as commonplace to teachers as using the chalkboard and overhead projector. Fourth, teachers must establish appropriate instructional goals and inquiry strategies and be able to guide students' learning accordingly.

The instructional use of technology must reinforce students' learning. The emphasis should be on analytical problem-solving skills. The focus must be on concept development, not merely on procedures with limited use outside the classroom (e.g., computation). Computer technology must be fully available to *all* students and teachers at *all* times. Since computers are being integrated into every aspect of the workplace, the integration of technology into teaching and learning is essential. Teaching and learning require the careful sequencing of activities, not just "topic after topic" sequencing in isolation with little benefit to students.

CONCLUSION

The use of computer technology is an integral part of our vision of the future of the teaching and learning of mathematics. The "thinking skills" involved in the software that we have identified include making inferences, predicting events, recognizing patterns, grouping events, formulating hypotheses, analyzing data, making generalizations, evaluating outcomes, classifying objects, and controlling variables. In every case, curriculum objectives must be matched with the context of mathematical situations, methodology, and resources.

If past history is a valid base for predicting the future, methods of teaching mathematics in the 1990s will be little different from those used in the past forty years. Despite the broad use of computer technology outside education, extensive societal and technological change, and the prognostications of many groups, mathematics instruction has changed little. The time has come to provide mathematical instruction in a way that is consistent with advances in technology and society and in ways that have not previously been possible. Teachers have the opportunity to reconceptualize their role in the classroom and have an obligation to involve students more actively in learning. Technological aids allow greater realism in the classroom, which in turn calls for reexamining the *context* of teaching and learning of mathematics.

REFERENCES

Bollinger, Galen. "muMATH and High School Math." Mimeographed. Austin: Department of Curriculum and Instruction, College of Education, University of Texas at Austin, 1986.

Cole, Michael, and Peg Griffin, eds. *Improving Science and Mathematics Education for Minorities and Women: Contextual Factors in Education*. Madison: Wisconsin Center for Education Research, 1987.

Palmiter, Jeanette R. "The Impact of a Computer Algebra System on College Calculus." *Dissertation Abstracts International* 47 (1986): 1640A.

27

Part 7

Mathematics Teachers Reconceptualizing Their Roles

Charles Lovitt
Max Stephens
Doug Clarke
Thomas A. Romberg

REFORM documents from several countries have argued for far-reaching changes in the teaching of school mathematics. These proposals, while detailing the pressures and directions for change, often do not elaborate on how changed practice is likely to come about or how teachers in ordinary classroom settings can become partners in changing their own practice.

As a profession, we have often neglected to identify the best current practice and to document it in such a way that the collective wisdom of our best teachers can help others to reflect on, and reshape their own teaching. Meeting this challenge is at the heart of this article and the program on which it is based.

INTRODUCTION

An agenda for the reform of school mathematics has been articulated with remarkable consistency by such varied documents as the NCTM's *Curriculum and Evaluation Standards for School Mathematics* (National Council of Teachers of Mathematics 1989), England's *National Criteria for Mathematics in the General Certificate of Secondary Education* (Department of Education and Science 1985), and an Australian *Mathematics Framework, Preparatory Year to Year 10* (Victorian Ministry of Education 1988). It is absolutely clear that these documents go far beyond a call for changes in the content of school mathematics. All have significant implications for teaching at every level.

The *Mathematics Framework (P–10)* for schools in one Australian state provides an important insight into this agenda for reform:

> There is now a constant and concerted call to improve mathematics teaching. . . . In all of this there can be a tendency to downgrade current practice.
>
> Yet current practice is the base we all build on and it contains much that is worthwhile. It embodies our store of mathematical understandings, and our knowledge of students and how they learn. (P.15)

Teachers of mathematics know from experience that reform documents and agendas for change do not of themselves bring about change in teaching. The overriding question for this article is how ideal images of mathematics teaching can be translated into practice in ordinary classroom settings. We shall describe elementary and secondary school mathematics teachers engaged in changing, expanding, and improving their classroom practice. We shall focus on processes that are proving to be successful in the Australian Mathematics Curriculum and Teaching Program (MCTP). This national program is a collaborative venture jointly sponsored by all Australian states and territories and follows a model established by the earlier Reality in Mathematics Education Project (RIME).

THREE PRINCIPLES FOR SUCCESSFUL REFORM

In our experience, three principles are fundamental to any reform of mathematics teaching and learning. These principles have been embedded in the MCTP. We see them as essential for bringing to life, in ordinary classroom settings, ideal images of mathematics teaching and learning.

1. Any program that seeks to enhance the quality of teaching and learning in mathematics must allow teachers to develop, in practical terms, a clear vision of what these changes mean for their own personal professional behavior. It implies that teachers actively reflect on their current practice and make a professional commitment to work toward an improved and expanded repertoire of teaching skills.
2. Exemplary curriculum materials can help teachers think about their current roles, try out new roles, and modify the way they teach by drawing directly on the accumulated experience of teachers who have helped to develop and try out these materials. Such materials are "exemplary" in the sense that they reflect what Shulman (1987) calls "the wisdom of practice."
3. Reshaping the teaching of mathematics requires that teachers have access to a sustaining and well-structured environment for their professional growth. We see no other way of working with teachers as they endeavor to bring about quality mathematics teaching in their classrooms.

As we explore each of these principles in greater detail, there is no implied sequence in any professional program designed to address them.

All three are intimately interwoven; any program should consider all three concurrently.

Principle 1: Develop a clarity of vision.

Any program that seeks to enhance the quality of teaching and learning in mathematics must allow teachers to develop, in practical terms, a clear vision of what these changes mean for their classroom practice and professional growth.

Central to this principle is the concept of the individual teacher recognizing and reflecting on his or her own personal repertoire of understandings about mathematics teaching and learning. The richer this repertoire, the greater the quality of the learning environments that the teacher can offer to pupils. This repertoire, which might be termed a teacher's "comfort zone," is unique to each individual and has fluid boundaries that vary with those intangible qualities of experience and confidence. Many teachers are seeking to improve and expand the quality and range of their current practice. This entails becoming aware of and exploring new approaches to teaching and learning and new areas of content, which may for a particular teacher lie outside his or her confidence or comfort zone.

Reform documents promote a relatively large list of features, which include the ability to effectively use group investigations, problem-solving approaches, estimation, visual imagery, technology, concrete aids, and an activity style of learning; to incorporate new content and equity concerns; to use mathematical modeling and applications; to recognize, value, and build on the real world of students; and to use a wide range of assessment techniques.

Lampert (1988) argues that any teacher in command of all these features would indeed be the ideal teacher envisaged by the reform documents. But realistically for many teachers, this rather imposing list of features lies outside their current practice or comfort zone. If, for example, quality mathematics learning requires a stronger emphasis on students working cooperatively in groups, how do teachers, for whom such an idea is *new educational territory,* get the practical experience on which they can judge the value and desirability of this feature? For many teachers, areas of mathematics such as statistics and discrete mathematics are currently outside their comfort zone.

It is absolutely critical that teachers be given opportunities to develop confidence in these new areas of school mathematics, and to establish a vision for an expanded and higher-quality repertoire. We now consider how all this can be accomplished and, once the vision is established, how current practice can be gradually reshaped.

Principle 2: Share the wisdom of practice.

Using exemplary materials, illustrating quality teaching, and drawing on the experiences and accumulated wisdom of our very best teachers can provide the basis on which we can all reflect and learn.

What are exemplary materials? If we could all observe the mathematics classrooms of our schools, would we all agree on those that deserve the title of "quality"? And if we do find one—a classroom environment we agree is exemplary, where a teacher has successfully put into practice some of those features advocated in the reform agendas—how would we document it? How do we capture its spirit and substance so that others can reflect on it and learn from it?

In documenting quality practice, Shulman (1987) argues that one particular component of the art of teaching, which he describes as pedagogical reasoning, is the least recognized and codified. Pedagogical reasoning is the "intellectualization," or deep thinking, of what good teachers do and why they do it. It might well be called the "wisdom of practice." That this wisdom has not been recorded leads Shulman to make a telling point:

> One of the frustrations of teaching as an occupation is its extensive individual and collective amnesia, the consistency with which the best creations of its practitioners are lost to both contemporary practitioners and future peers. (Pp. 11–12)

One failed attempt to document specific illustrations of quality teaching occurred early in the RIME project (1984). Thirty-five teachers were invited to "bring along their most proven, successful activity." They were then challenged to write or document the activity so that others might be able to repeat the success in their own classrooms. The attempt failed, not because the activities were not educationally sound, but because the documentation failed to incorporate all the ingredients necessary for success, particularly the ingredient Shulman calls pedagogical reasoning. Teachers wrote down *what* they did but not *why* they did it.

One promising style of documentation appears in the MCTP activity "This Goes with This." Alongside a running commentary of an illustrated activity in a classroom are anecdotes and insights into the "background" thinking of the teacher. The Mathematics Curriculum and Teaching Project (1988) has collected about 100 such illustrations into an activity bank. Does such documentation prove beneficial to others? Results thus far indicate strongly that transfer does take place and that teachers are gaining access to the pedagogical reasoning of our best teachers and in this manner are able to reshape and improve their own practice.

Specific classroom activities collected and documented in this way represent single "snapshot" images of quality learning environments. The aggregation of these images forms a gallery of the art of good teaching. Exemplary classroom activities presented in this manner should be intended as vehicles toward new understandings and reshaped practice, not ends in themselves.

The documentation of these exemplary activities and their subsequent discussion by teachers enable them to recognize and overcome the impediments and obstacles that so often stifle innovation. Otherwise, there is a real risk that initial lack of success with a new approach may lead to its rejection.

What the reform documents tend to pass over is that strategies for change must confront an already established culture. Teachers and students already hold strong beliefs about mathematics and how it is taught and learned. Romberg and Price (1983) argue that these beliefs and practices are often impediments to change. In a related study, Stephens and Romberg (1985) highlighted one of these impediments, namely, students' expectations. All the teachers interviewed drew attention to the importance of gaining students' acceptance of what is going to take place: what the mathematics lesson is about, what they will get out of it, and how it may be different from what they have done before. Teachers offered the following advice (Stephens and Romberg 1985, p. 5):

> Not to do so is to sow the seeds of destruction . . . they will reject it because it is different.
>
> No matter how philosophically sound a lesson may be, if it doesn't meet students' expectations about what a regular maths lesson is going to be, they become suspicious of it. If there is a shift in the way mathematics is going to be taught, then there must be something built into the lesson which takes account of students' expectations, and how they might be changed.
>
> Sometimes a lesson turns out to be a disaster for a whole lot of little reasons. Not the least of these reasons is students' expectations of what a real mathematics lesson is supposed to be about.

Principle 3: Provide in-service models to facilitate change.

A structured, supportive professional-development environment must be provided in order for teachers to reshape their practice.

Access to exemplary materials can help teachers to develop a clear image of directions for self-improvement, but a supporting environment in which the consideration and exploration of the materials takes place is necessary for lasting change.

There is no single ideal approach; rather, we have found a range of approaches or models by which teachers might be assisted. The following models are a synthesis of good professional-development practice drawn from theoretical perspectives and practical experience; they represent the best we have been able to find.

It will be obvious from what follows that the "one shot" in-service program is not included. This approach is in conflict with many established principles of professional development. It may be a useful means of transmitting information, but it is unlikely to lead to lasting change. Although the models may appear different in organizational detail, they are underpinned by the following set of key features identified by research and our experience. To be effective, a professional-development program should—

- address issues of concern recognized by the teachers themselves;
- be as close as possible to the teacher's working environment;
- take place over an extended period of time;
- have the support of both teachers and the school administration;
- provide opportunities for reflection and feedback;
- enable participating teachers to feel a substantial degree of ownership;
- involve a conscious commitment on the part of the teacher;
- involve groups of teachers rather than individuals from a school;
- use the services of a consultant or critical friend.

We have identified and documented eight generic types of models (see Owen et al. 1988), each of which embodies to varying degrees the features listed above. These generic models are the *structured course, the sandwich model, in-school intensive, school cluster group, postal model, a preservice course, peer tutoring,* and *activity documentation.* Two of these are now discussed in greater detail.

The Sandwich Model

The sandwich model has the following components:

- Two seminar sessions take place two to three weeks apart.
- Classroom trials are "sandwiched" between these two sessions.
- The focus is usually on only one selected theme.
- Session 1 is a workshop on exemplary activities on this theme.
- Session 2 discusses the classroom trials and plans further actions at the school level.

This model allows a group of teachers to explore new approaches that they have selected and see as desirable. An exemplary activity is presented in the first session, ideally by demonstration in a classroom. In this session, the mathematical and pedagogical purposes of the activity need to be identified. The follow-up session provides a critical opportunity for teachers to report on their trials of the activity. They should discuss unexpected variations, impediments, and possible extensions and relate these to the pedagogical and mathematical purposes. This session also provides a basis for planning future action.

For these reasons, the sandwich model is a vast improvement over the one-shot approach. It is a powerful means of introducing an innovation to teachers. Other advantages of sandwich courses are that they are easy to set up, are inexpensive, and are readily embraced by busy teachers. Although full-day meetings are preferable, two-hour meetings after school can operate successfully.

The Activity Documentation Model

This type of approach has the following structure:

- Teachers bring notes and equipment for one activity that has resulted in significant success in their school.
- These activities are illustrated and shared with small groups, and initial documentation is developed.
- Teachers undertake to try out at least two of the activities.
- Between-session trials occur.
- A second meeting focuses on further documentation of the activities.
- These activities and results are circulated to the total group and any interested others.

As the name implies, this model involves the documentation of successful innovative practice so that exemplary activities developed by one teacher or a team of teachers can be more widely circulated. Many teachers are working at the "cutting edge" of innovative mathematics teaching and are successfully implementing new dimensions into their teaching. Tapping into their knowledge (or "teaching wisdom") is the purpose of activity documentation.

Trying to capture the spirit as well as the substance of a successful activity—especially its pedagogical and mathematical structure—is a complex and difficult task. The documentation needs to function as a window into the mind of the teacher and contain *all* the insights, anecdotes, hints, and pedagogical reasoning that generated the success. After trying, discussing, and rigorously analyzing quality activities, all the teachers involved in these models recognized that the success lies not so much in the mathematics but in recognizing what teachers do and why they do it. Hence, the documentation must include all those small, important, and illuminating details that often go unnoticed and unshared and that yet collectively hold the secrets to success.

CONCLUSION

Unless an agenda for the reform of school mathematics respects the principles advocated in this article, there is a real danger that the recommended changes will once again be seen by teachers as imposed from above. Teachers need to know where they are going and why they are going there. They need to have genuine access to, and to share, the considerable "wisdom of practice" that already exists in their classrooms. They also need to be supported in their growth by theoretically sound and supportive environments. In these ways, teachers can be active partners in reconceptualizing their roles.

REFERENCES

Department of Education and Science. *National Criteria for Mathematics in the General Certificate of Secondary Education.* London: Her Majesty's Stationery Office, 1985.

Lampert, Magdalene. "What Can Research on Teacher Education Tell Us about Improving Quality in Mathematics Education?" *Teaching and Teacher Education* 4 (1988): 157–70.

Mathematics Curriculum and Teaching Program. *Professional Development Package.* Canberra, Australia: Curriculum Development Centre, 1988.

National Council of Teachers of Mathematics. *Curriculum and Evaluation Standards for School Mathematics.* Reston, Va.: The Council, 1989.

Owen, John, Neville Johnson, Doug Clarke, Charles Lovitt, and Will Morony. *Professional Development in the Mathematics Curriculum and Teaching Program: Guidelines for Consultants and Advisors.* Canberra, Australia: Curriculum Development Centre, 1988.

Reality in Mathematics Education (RIME). *Teacher Development Pack.* Melbourne, Australia: Victorian Ministry of Education, Curriculum Branch, 1984.

Romberg, Thomas A., and Gary Price. "Curriculum Implementation and Staff Development as Cultural Change." In *Staff Development,* 82d Yearbook of the National Society for the Study of Education, edited by Gary A. Griffin, pp. 154–84. Chicago: University of Chicago Press, 1983.

Shulman, Lee S. "Knowledge and Teaching: Foundations of the New Reform." *Harvard Educational Review* 7(1) (1987): 1–22.

Stephens, W. M., and Thomas A. Romberg. "Reconceptualizing the Role of the Mathematics Teacher." Paper presented at the Annual Meeting of the American Educational Research Association, Chicago, 1985.

Victorian Ministry of Education. Schools Division. *Mathematics Framework: P–10, a Forward Look.* Victoria, Australia: The Ministry, 1988.

28

Professionals in a Changing Profession

Mark Driscoll
Brian Lord

THIS yearbook serves a dual purpose. It is primarily a look to the future, to mathematics education in the next decade. But it is also a look back to the decade just ending; in particular, it is a reflection of the sweeping changes of the 1980s—in mathematics, technology, and education research—that have paved the way for a renewed field of mathematics education in the 1990s.

There is one movement toward change, however, that has not been so visible within the education community, and it is the focus of this article. It concerns the professional lives of mathematics teachers. The 1980s saw the beginning of several efforts to reshape the mathematics teaching profession, and if those efforts continue to expand and thrive, mathematics teachers in the 1990s will have the option of assuming roles and responsibilities that are more varied, more challenging, and potentially more satisfying than those defining the profession today.

We shall address the changing roles and responsibilities of mathematics teachers (1) in the classroom, (2) in the teaching profession, and (3) in the broader community. Since policy decisions at local and state levels heavily affect possibilities for change within the teaching profession, we shall concentrate in the last section on policy changes that will be necessary in order for enhanced roles and responsibilities to be made available to mathematics teachers in the 1990s.

CHANGING ROLES AND RESPONSIBILITIES: IN THE CLASSROOM

As we enter the 1990s, it remains to be seen whether a major shift will occur in how mathematics is taught as suggested by those advocating reform in mathematics education. It is undeniable, however, that forces abound that suggest change in mathematics education is inevitable. For example, mathematicians for over a decade have used the computer to surge ahead in their research, yet the curriculum has resisted computer tools and the exciting mathematics topics they make accessible (Hilton 1984). A stall, if not an actual decline, in the mathematics achievement of American learners has haunted us in recent international comparisons (Steen 1987)—an especially frustrating pattern, since researchers in the past decade have pinpointed numerous components of effective mathematics teaching (Grouws and Cooney 1988).

These conditions have not appeared in a vacuum. They reflect a rapidly changing world that both surrounds and influences mathematics teachers and learners. In the past couple of decades, the meaning of "mathematically literate" has been transformed by developments in mathematics, technology, and learning theory. The effect on schooling is to make us look again at the roles and responsibilities of mathematics teachers in the classroom.

How might those roles and responsibilities change? In brief, they will change according to major shifts in mathematics, technology, and cognitive science. Some of the shifts in these fields have been explored in other articles in this yearbook. Mathematical models that use discrete methods and methods of approximation have proliferated in the world of mathematics. The calculator and computer have become essential tools for anyone who wishes to explore such models; in addition, they make such exploration accessible to students at all school levels. Mathematics educators and psychologists have come to recognize that mathematics learning is more than the absorption of information, that learners should be actively engaged by teachers to weave the information into their observations and interpretations of the world around them.

To keep pace with these changes, mathematics teachers must alter the ways in which they perceive the field of mathematics and the processes of teaching and learning: it is an open-ended field, enriched by exploration and experimentation, and is not a static body of facts. Teaching mathematics should reflect the dynamic, experimental quality of the discipline and so should be much more than merely delivering information. In particular, teaching mathematics should involve the technological tools that foster experimentation. Finally, learning mathematics is inseparable from doing mathematics, even at the most elementary levels, and so the mathematics classroom should be a place where learners are encouraged not only to learn the appropriate facts and skills but also to *do* mathematics and to reveal the thinking behind their mathematical activity.

Research on the teaching of mathematics has a history of being guided by the "nutrition" metaphor for teaching (Schon 1983). According to this metaphor, "Teachers are seen as technical experts who impart privileged knowledge to students. . . . Children are fed portions of knowledge, in measured doses. They are expected to digest it and to give evidence, in class response and examinations, that they have done so. The curriculum is conceived as a menu of information and skills, each lesson plan is a serving, and the entire process is treated as a cumulative, progressive development" (Schon 1983, p. 329).

Metaphorically speaking, the mathematics teacher ought to be less of a nutritionist in instruction, and more of a guide, coach, and psychologist. The teacher as guide leads students into important areas for investigation and learning through a pattern of appropriate modeling and of probing and challenging questions. The teacher as coach raises all students beyond their too often low expectations for success and inspires them to use that oft-neglected component of good problem solving: persistence. The teacher as psychologist recognizes, in the words of Brown (1985), that "regardless of age and level of sophistication of our students, they have attitudes, values, beliefs, and feelings which form a window that affects how they see the world" (p. 22). Emotional growth and intellectual growth are inseparably intertwined in the mathematics classroom, and classroom instruction should reflect this fact.

CHANGING ROLES AND RESPONSIBILITIES: IN THE PROFESSION

As the decade of the 1980s comes to a close, there are signs that the mathematics teaching profession is moving away from the status to which history has consigned and confined it. Schools in America have evolved on what Dan Lortie has called "the egg-crate model" (Lortie 1975). As the country grew rapidly in the late nineteenth century, education's needs quickly outgrew the one-room schoolhouse. However, as Lortie describes in his book *Schoolteacher,* the vast majority of teachers and teacher candidates were either married men who felt destined for administrative positions or single women who routinely were expected to marry within a short time. (Married women were not allowed to teach.) The combined effect conveyed the impression that teaching was a highly transient profession and one where teamwork was unlikely to develop. Hence the egg-crate metaphor: with the turnover as high as it was—and was expected to be—it was convenient to design the typical school as if it were a collection of one-room schoolhouses. Teachers were separated from one another, and collegiality was not encouraged. Thus a basic component of a healthy profession was stifled from the start.

As a rule, teachers are still separated from one another professionally. As one consequence, they have become dependent on the decisions of outside experts—education researchers, administrators, curriculum developers, textbook writers, and test designers—who set the standards by which classroom instruction is judged. This division of labor exhibited in the processes and products of instruction has resulted in the "de-skilling" of the teaching profession (Apple 1982).

During the past few decades, mathematics education has built a knowledge base that has made substantial inroads into replacing opinion as the foundation of mathematics teaching. Unfortunately, in defining the professional roles and responsibilities of mathematics teachers, we have come to rely too much on a static application of theory to practice. Consequently, teachers are judged by criteria that ignore the dynamic qualities of good instruction. For example, successful mathematics teachers use a rule-of-thumb knowledge in their day-to-day work. This is the sort of knowledge that leads them to recognize and make use of such "teachable moments" as when students ask "what if . . ." questions or when they reveal misconceptions about mathematical concepts. It is the sort of knowledge that teachers use to risk confusion and to welcome surprise in their interactions with students. These patterns of teaching are representative of the kind of knowledge Schon calls "knowledge-in-action," a series of actions, recognitions, and judgments that successful professionals in a variety of fields learn to carry out in the course of their work, especially when problems arise that do not fit existing rules (Schon 1983). Teaching—and mathematics teaching, in particular—is too complex, unstable, and laced with uncertainty to be described and evaluated primarily by static, cookbook criteria. There needs to be more understanding of, and reliance on, knowledge-in-action among mathematics teachers. Unfortunately, the egg-crate design of their profession has kept them from reflecting adequately on their knowledge-in-action and from sharing it with one another. Therefore, much of teachers' best knowledge lies hidden and unused.

As we look ahead, however, to the roles and responsibilities of mathematics teachers in the 1990s, there is hope that the professional isolation of mathematics teachers will subside and that they will become better able to take advantage of their knowledge-in-action. This hope derives from efforts of the last decade and signs that those efforts will continue to bear fruit.

The Study of Exemplary Mathematics Programs of the early 1980s pointed to the importance of staff collegiality in the success of school mathematics programs. In the majority of the exemplary programs studied, mathematics teachers shared with one another, worked toward cohesion in their planning and implementation, and often took risks together (Driscoll 1986).

Also during the early eighties, Little (1982) conducted a study in which she compared classrooms in schools where collegiality was part of the teachers' way of life to classrooms in schools where teachers worked in relative isolation. Her working definition of *collegiality* comprised four behaviors: teachers talk about the practice of teaching and learning frequently; they observe one another; they work on the curriculum together; and they teach one another what they know about teaching, learning, and leading. Just as in the Study of Exemplary Mathematics Programs, Little found similar correlations of collegiality with successful student outcomes.

In the middle to latter part of the eighties, several projects arose to help mathematics teachers break their professional isolation, to establish mechanisms by which they could sustain ongoing dialogues with one another, and to develop the leadership skills necessary to shape the profession so that they are the principal experts and solvers of problems that arise within the profession.

The California Mathematics Project (CMP) (n.d.) serves sixteen sites around the state and offers programs for teachers interested in leadership roles in mathematics education. There are summer institutes and follow-up activities throughout the year that promote the leadership of teachers in solving some of the major problems facing the profession (e.g., making mathematics more accessible to minorities and women).

CMP is just one of the new projects that envision mathematics teachers as critical, reflective professionals who engage in collegial problem solving with their peers and with others concerned about mathematics education. Another is the Project to Increase Mastery of Mathematics and Science (PIMMS), a project in Connecticut with similar goals to those of CMP.

A third project that set out in the decade of the eighties to open and enrich the mathematics teaching profession is the Urban Mathematics Collaboratives (UMC) Project. The UMC Project, initiated by the Ford Foundation, is designed to improve mathematics education in inner-city schools by making it possible for urban mathematics teachers in eleven large American cities to define new models for meeting their ongoing professional needs (Ford Foundation 1987).

The collaboratives aspire to make it possible for mathematics teachers in each of the eleven UMC cities to influence the evolution of local visions of mathematics education and the development of policies on which those visions hinge. In varying proportions across the UMC network, the collaboratives sponsor workshop series and seminars; internship-in-industry programs; departmental leadership training; teacher-run conferences; computer conferences around topics of mutual interest (e.g., teaching geometry using computers); symposia with business and academic colleagues to discuss pending reform in mathematics education; and teacher-run research projects.

The evaluation of the effectiveness of the collaboratives is incomplete, but they appear to be fostering professional change in several ways. In particular, when teachers engage in an ongoing workshop series or communicate regularly in an electronic conference that is focused on a topic of mutual interest, they are more inclined to reflect on, and make visible, their knowledge-in-action as mathematics teachers. Thus, they are better able to bring their best knowledge to the classroom.

CHANGING ROLES AND RESPONSIBILITIES: THE BROADER COMMUNITY

In the preceding section, we discussed the importance of overcoming teacher isolation and observed that recent efforts to improve mathematics education have focused on teachers' collegiality and dialogue. Implicit in such efforts and in the collegial ties they promote is an exciting new vision of the teaching profession. This vision regards relationships among colleagues as *constitutive* of a profession and not merely as consequences or rewards of a profession. It is through these relationships, both at the core and at the periphery of a discipline, that the language of a profession is clarified and its problems posed. If professional practice is an inherently social activity (and one might argue that it is, requiring at the very least a practitioner and a client), it is even more so in the case of teaching, which traces its roots to the fundamentally public and democratic objectives of schooling. Teaching, as a profession, will be shaped, in substantial part, by its relationship to the social, economic, and political institutions that move the wider society. This *relational* view of professionalism takes seriously the need for change and growth in teachers' knowledge (about content, about pedagogy, and about student learning) and locates the source of this growth in a broader network of professional resources.

This point is especially instructive for understanding the changing roles and responsibilities of mathematics teachers in the broader community. If mathematics teachers are to be received as professionals by this community, then they must assume responsibility for developing professional relationships with this community. By working more closely with representatives of business, academic, cultural, and political institutions, the mathematics teachers of the 1990s will have opportunities to overcome teacher isolation and reinvigorate mathematics education as well as themselves. It will not be enough, however, to wait for resources and support from outside agents committed to change; instead, teachers must think and act as part of a community that makes use of mathematics, a community that has substantial investment in securing this change. Teachers must take steps to build enduring collegial ties both within and beyond the school and assume leadership on the pressing issues of curriculum reform and professional renewal.

How can mathematics teachers foster these broader networks, giving shape to their profession while nurturing a dialogue in support of change? One way, mentioned earlier, is through teacher professionalism projects or collaboratives (e.g., the California Math Project, PIMMS, or the Urban

Mathematics Collaboratives). An emerging network of these projects has sought to initiate and sustain effective linkages between mathematics teachers and the wider community of mathematics professionals. In many cities, teachers have begun working with representatives of business, academic, and cultural institutions to find new and more interactive strategies for promoting professional development and improving mathematics instruction. Whether through workshops, seminars, dinner speakers' engagements, summer institutes, professional conferences, or internships in industry, mathematics teachers have embraced collaboration as a way to broaden the base of institutional support, enlarge the conversation about mathematics teaching and learning, and further legitimize their standing in the community of mathematics professionals. In effect, they have initiated an ever more stimulating and responsive "community of inquiry" around the concerns of mathematics education, a powerful catalyst for professional development and curriculum change (Gaudiani and Burnett 1985, p. 5).

Yet another role that mathematics teachers might play in their efforts to build relationships with the broader community is that of "designer" or "critic" of reform in mathematics education and of education reform more broadly. Wise (1987) describes a professional as one who "must make decisions on behalf of less knowledgeable clients in settings where no 'higher authority' (except professional ethics) is present" (p. 2). Through critical reflection on the proposed changes in how mathematics is taught (e.g., in the NCTM's *Curriculum and Evaluation Standards for School Mathematics* or in MSEB's Framework for School Mathematics), many teachers are developing skills crucial to the kind of decision making that Wise describes. By focusing on the strengths and weaknesses in the content of what they teach, mathematics teachers are better prepared to exercise professional judgment and make decisions in the best interests of their students.

Teachers who actively participate in professional networks or engage in critical reflection on their discipline soon recognize that education systems can present structural obstacles to professional growth and change. They begin to ask broader questions about education reform, about ways to restructure authority relationships, school organization, and student assessment. In short, they seek roles as designers and critics of the wider reform movement; they try to devise openings through which they can introduce innovation and sound judgment into the art and craft of teaching. Elmore and McLaughlin (1988, p. 8) rightly observe that

> if teachers have thought about their own view of practice and if they have strong professional convictions about how to teach effectively, they are more likely to see conflicts between reform policies and their own work. Teachers are often the last to be heard from on the effects of reform policies and the first to be criticized when reforms fail.

If teachers are to forestall unwarranted criticism, then they must assume responsibility for exposing the conflicts between reform policies and the requirements of their own work. More important, they must acquire and exercise the skills of political advocacy that are among the hallmarks of professional life. Mathematics teachers have already begun to assume these new roles and to establish new relationships with the broader education community.

CONCLUSION

The kind of "revolution" in mathematics education described in this article and elsewhere in this yearbook will be strongly influenced by the nation's broader agenda for education reform in the 1990s. As we move into a new decade, the climate for structural reform of schooling and, in particular, for accelerated change in the organization of the teaching profession is favorable. The principal impact of the "first wave of reform," from its inception in the national reports of the early 1980s to its culmination in the omnibus education initiatives enacted by a majority of states, has been to focus attention on the conditions of teaching and on acceptable standards for entering and remaining in the profession.

The problems that inspired this initial spate of reform are still with us, however, and will not easily be solved. Impending teacher shortages (regionally and in specific subject areas), the technological and social demands of a rapidly changing economy, and the burgeoning population of "at risk" students for whom the education system has been largely ineffective have all led to calls for the wholesale "restructuring" of the teaching profession. Teachers, the argument goes, must be allowed more discretion in the classroom, more decision-making powers within the school, expanded opportunities for collegial exchange, increased salaries, and more realistic career paths if they are to assume positions of leadership in the nation's drive toward improvement in education.

Even though greater autonomy, more realistic rewards, and increased opportunities for professional renewal are all necessary conditions for change, these reforms may not be sufficient for change. If "restructuring" is to work, and it must if we are to realize many of the goals we have set for mathematics teaching and learning, teachers must take hold of these organizational changes to restructure professional relationships among themselves and between themselves and others having a stake in the education system—students, parents, administrators, policymakers, and the wider public. Ultimately, the attractiveness of the profession and the extent to which it can embrace important new curricular and instructional changes will hinge on how effectively teachers capitalize on the intellectual and organizational possibilities implicit in these new professional environments.

The challenge of professional autonomy, as McNeil (1987) observes, is one of "empowering and enabling teachers, so that they will feel confident to bring their best knowledge into schools" (p. 4). The nation's mathematics teachers have already made important inroads into structuring an autonomous profession and into using that autonomy to enlarge the conversation about mathematical knowledge and mathematical learning. As the nation's policymakers move toward building a broader base for this kind of reform, mathematics educators can look forward to a continuing revolution in the decade ahead.

REFERENCES

Apple, Michael W. *Education and Power*. Boston: Routledge & Kegan Paul, 1982.

Brown, Stephen. "Problem Solving and Teacher Education: The Humanism Twixt Models and Muddles." In *Studies in Mathematics Education: The Education of Secondary School Teachers of Mathematics*, vol. 4, edited by Robert Morris. Paris: UNESCO, 1985.

California Mathematics Project. Berkeley, Calif.: Office of the President, University of California, Berkeley, n.d.

Driscoll, Mark. *Stories of Excellence: Ten Case Studies from A Study of Exemplary Mathematics Programs*. Reston, Va.: National Council of Teachers of Mathematics, 1986.

Elmore, Richard F., and Milbrey Wallin McLaughlin. *Steady Work: Policy, Practice, and the Reform of American Education*. Santa Monica, Calif.: Rand Corp., 1988.

Ford Foundation. . . . *And Gladly Teach*. New York: The Foundation, 1987.

Gaudiani, Claire L., and David G. Burnett. *Academic Alliances: A New Approach to School/College Collaboration*. Washington, D.C.: American Association for Higher Education, 1985–86.

Grouws, Douglas A., and Thomas J. Cooney, eds. *Effective Mathematics Teaching*. Reston, Va.: National Council of Teachers of Mathematics, 1988.

Hilton, Peter. "Current Trends in Mathematics and Future Trends in Mathematics Education." *For the Learning of Mathematics* 4 (February 1984): 2–8.

Little, Judith Warren. "Norms of Collegiality and Experimentation: Workplace Conditions of School Success." *American Educational Research Journal* 19 (Fall 1982): 325–40.

Lortie, Dan C. *Schoolteacher*. Chicago: University of Chicago Press, 1975.

McNeil, Linda. *Proceedings of the Third Annual Meeting of the Urban Mathematics Collaboratives*. Newton, Mass.: Education Development Center, 1987.

Schon, Donald A. *The Reflective Practitioner: How Professionals Think in Action*. New York: Basic Books, 1983.

Steen, Lynn Arthur. "Mathematics Education: A Predictor of Scientific Competitiveness." *Science*, 17 July 1987.

Wise, Arthur E. *Annual Report of the Center for the Study of the Teaching Profession, 1986–87*. Santa Monica, Calif.: Rand Corp., 1987.

Index